Gerhard • Oevel • Postel • l
MuPAD Tutorial

Springer

*Berlin
Heidelberg
New York
Barcelona
Hong Kong
London
Milan
Paris
Singapore
Tokyo*

J. Gerhard • W. Oevel • F. Postel • S. Wehmeier

MuPAD Tutorial

English Edition
A version and platform independent introduction

 Springer

Jürgen Gerhard
Walter Oevel
SciFace Software GmbH & Co. KG
Technologiepark 11
33100 Paderborn, Germany

Frank Postel
Stefan Wehmeier
Universität–GH Paderborn
FB17 Mathematik/Informatik
Warburger Str. 100
33098 Paderborn, Germany

CIP data applied for
Die Deutsche Bibliothek - CIP-Einheitsaufnahme

MuPAD tutorial: a version and platform independent introduction / J. Gerhard... - English ed.-
Berlin; Heidelberg; New York; Barcelona; Hong Kong; London; Milan; Paris; Singapore; Tokyo:
Springer, 2000
Dt. Ausg. u. d. T.: Das MuPAD-Tutorium
ISBN 978-3-540-67546-4   ISBN 978-3-642-98114-2 (eBook)
DOI: 10.1007/978-3-642-98114-2

Mathematics Subject Classification (1991): 68Q40

ISBN 978-3-540-67546-4

Springer-Verlag Berlin Heidelberg New York
a member of BertelsmannSpringer Science+Business Media GmbH

© Springer-Verlag Berlin Heidelberg 2000

The use of general descriptive names, registered names, trademarks etc. in this publication does not
imply, even in the absence of a specific statement, that such names are exempt from the relevant
protective laws and regulations and therefore free for general use.

Cover design: *design & production GmbH*, Heidelberg
Typesetting by the authors using a Springer LaTeX macro package
Printed on acid-free paper   SPIN 10764648   40/3143ck-5 4 3 2 1 0

Preface

This book explains the basic use of the software package called MuPAD and gives an insight into the power of the system. MuPAD is a so-called computer algebra system, which is developed mainly at the University of Paderborn in Germany.

This introduction addresses mathematicians, engineers, computer scientists, natural scientists, and more generally all those in need of mathematical computations for their education or their profession. Generally speaking, this book addresses anybody who wants to use the power of a modern computer algebra package.

There are two ways to use a computer algebra system. On the one hand, you may use the mathematical knowledge it incorporates by calling system functions interactively. For example, you can compute symbolic integrals, or generate and invert matrices, by calling appropriate functions. They comprise the system's mathematical intelligence and may implement sophisticated algorithms. Chapters 2 through 15 discuss this way of using MuPAD.

On the other hand, with the help of MuPAD's programming language you can easily add functionality to the system by implementing your own algorithms as MuPAD procedures. This is useful for special purpose applications if no appropriate system functions exist. Chapters 16 through 18 are an introduction to programming in MuPAD.

You can now read this book in the standard way "linearly" from the first to the last page. However, there are reasons to proceed otherwise. This may be the case, e.g., if you are interested in a particular problem, or if you already know something about MuPAD.

For MuPAD beginners, we recommend to start reading Chapter 2, which gives a first survey of MuPAD. The description of the online help system in Section 2.1 is probably the most important part of this book. The help system provides information about details of system functions, their syntax, their calling parameters, etc., and is available online during a MuPAD session. In the beginning, requesting a help page is probably your most frequent query to the system. After you have grown familiar with the help system, you may start to experiment with MuPAD. Chapter 2 demonstrates some of the most important system functions "at work". You find further details about these functions in later parts of the book or in the help pages. For a deeper understanding of the data structures involved, you may consult the corresponding sections in Chapter 4.

Chapter 3 discusses the MuPAD libraries and their use. They contain many functions and algorithms for particular mathematical topics.

The basic data types and the most important system functions for their manipulation are introduced in Chapter 4. It is not necessary to study all of them in the same depth. Depending on your intended application, you may selectively read only the passages about the relevant topics.

Chapter 5 explains how MuPAD evaluates objects; we strongly recommend to read this chapter.

Chapters 6 through 11 discuss the use of some particularly important system functions: substitution, differentiation, symbolic integration, equation solving, random number generation, and graphic commands.

Several useful features such as the history mechanism, input and output routines, or the definition of user preferences are described in Chapters 12 through 14. The preferences can be used to configure the system's interactive behavior after the user's fancy, to a certain extent.

Chapters 16 through 18 give an introduction to the basic concepts of the MuPAD programming language.

MuPAD provides algorithms which can handle a large class of mathematical objects and computational tasks related to them. If you work with this introduction, it is possible that you encounter unknown mathematical notions such as rings or fields. This introduction is not intended to explain the mathematical background for such objects. Basic mathematical knowledge is helpful but not mandatory to understand the text. Sometimes you may ask what algorithm MuPAD uses to solve a particular problem. The internal mode of operation of the MuPAD procedures is not addressed here: we do not intend to give a general introduction to computer algebra and its algorithms. The interested reader may consult text books such as, e.g., [GCL 92] or [GG 99].

This book gives an *elementary* introduction to MuPAD. Somewhat more abstract mathematical objects such as, e.g., field extensions, are easy to describe and to handle in MuPAD. However, such advanced aspects of the system are not discussed here. The mathematical applications that are mentioned in the text are intentionally kept on a rather elementary level. This is to keep this text plain for readers with little mathematical background and to make it applicable in school.

We cannot explain the complete functionality of MuPAD in this introduction. Some parts of the system are mentioned only briefly. It is beyond the scope of this tutorial to go into the details of the full power of MuPAD's programming language. You find these in the MuPAD User's Manual [MuP 96] and in MuPAD's help system. Both references are available online during a MuPAD session.

This tutorial refers to MuPAD versions 1.4 and later. Since the development of the system advances continuously, some of the details described may change sporadically in the future. Definitely, future versions provide additional functionality through new system functions and application packages. In transition to the currently developed version, the MuPAD kernel is subject to extensive changes. These in turn imply changes in the use of the programming language. In this tutorial, we mainly present the basic tools and their use, which remain essentially unchanged. We try to word all statements in the text in such a way that they stay basically valid for future MuPAD versions. The reader can download addenda and updates for this tutorial concerning MuPAD versions beyond 1.4 from the following web page:

http://www.sciface.com/support/springer

We would like to thank Tony Scott for proofreading the translation.

Paderborn, June 2000

Contents

1. Introduction

To explain the notion of computer algebra, we compare algebraic and numerical computations. Both kinds are supported by a computer, but there are fundamental differences, which are discussed in what follows.

1.1 Numerical Computation

A mathematical problem can be solved approximately by numerical computations. The computation steps operate on *numbers*, which are stored internally in *floating-point representation*. This representation has the drawback that neither computations nor solutions are exact, e.g., due to rounding errors. In general, numerical algorithms find approximate solutions as fast as possible. Often such solutions are the only way to handle a mathematical problem computationally, in particular if there is no "closed form" solution known. Moreover, approximate solutions are useful if exact results are unnecessary (e.g., in visualization).

1.2 Computer Algebra

In contrast to numerical computations, there are *symbolic* computations in computer algebra. [Hec 93] defines them as *"computations with symbols representing mathematical objects"*. Here, an *object* may be a number, but also a polynomial, an equation, an expression or a formula, a function, a group, a ring, or any other mathematical object. Symbolic computations with numbers are carried out *exactly*. Internally, numbers are represented as quotients of integers of arbitrary length (limited by the amount of storage available, of course). These kinds of computations are called *symbolic* or *algebraic*. [Hec 93] gives the following definitions:

1. "Symbolic" emphasizes that in many cases the ultimate goal of mathematical problem solving is expressing the answer in a closed formula or finding a symbolic approximation.
2. "Algebraic" means that computations are carried out exactly, according to the rules of algebra, instead of using the approximate floating-point arithmetic.

Sometimes "symbolic manipulation" or "formula manipulation" are used as synonyms for computer algebra, since computations operate on symbols and formulae. Examples are the symbolic integration or differentiation like

$$\int x \, dx = \frac{x^2}{2}, \quad \int_1^4 x \, dx = \frac{15}{2}, \quad \frac{d}{dx} \ln \ln x = \frac{1}{x \ln x}$$

or the symbolic solution of equations. For example, we consider the equation $x^4 + p \, x^2 + 1 = 0$ in x with one parameter p. Its solution set is

$$\left\{ \pm \frac{\sqrt{2} \, \sqrt{-p - \sqrt{p^2 - 4}}}{2}, \ \pm \frac{\sqrt{2} \, \sqrt{-p + \sqrt{p^2 - 4}}}{2} \right\}.$$

The symbolic computation of an exact solution usuallt requires more computing time and more storage than the numeric computation of an approximate solution. However, a symbolic solution is exact, more general, and often provides more information about the problem and its solution. The above formula expresses the solutions of the equation in terms of the parameter p. It shows how the solutions depend functionally on p. This information can be used, for example, to examine how sensitive the solutions behave when the parameter changes.

Combinations of symbolic and numeric methods are useful for special applications. For example, there are algorithms in computer algebra that benefit from efficient hardware floating-point arithmetic. On the other hand, it may be useful to simplify a problem from numerical analysis symbolically before applying the actual approximation algorithm.

1.3 Characteristics of Computer Algebra Systems

Most of the current computer algebra systems can be used interactively. The user enters some formulae and commands, and the system *evaluates* them. Then it returns an answer, which can be further manipulated afterwards if necessary.

In addition to exact symbolic computations, most computer algebra packages can approximate solutions numerically as well. The user can set the precision to the desired number of digits. In MuPAD, the global variable `DIGITS` handles this. For example, if you enter the simple command `DIGITS:=100`, then MuPAD performs all floating-point calculations with a precision of 100 decimal digits. Of course, such computations need more computing time and more storage than the use of the hardware floating-point arithmetic.

Moreover, modern computer algebra systems provide a powerful programming language[1] and tools for visualization and animation of mathematical data. Also, many systems can produce layouted documents (known as *notebooks* or *worksheets*). MuPAD has such a notebook concept, but we do not mention it in this tutorial. The goal of this book is to give an introduction to the mathematical power of MuPAD.

1.4 Existing Systems

There are many different computer algebra systems. Some of them are distributed commercially, while others can be obtained for free.

Special purpose systems can handle particular mathematical problems. For example, the system *Schoonship* is designed for problems in high energy physics, *DELiA* for differential equations, *PARI* for applications in number theory[2], and *GAP* for problems in group theory.

Then there are the *general purpose* computer algebra systems. *Derive* has been developed since 1980 and is designed specially for minicomputers. *MathView* (formerly *Theorist*), developed since 1990, has an elaborate user interface, but only comparably little mathematical knowledge. The systems *Macsyma* and *Reduce* have both started in 1965 and are programmed in LISP. Modern systems like *Mathematica* and *Maple*, developed since the beginning of the 80's, are programmed in C. *Mathematica* was the first system with a user friendly interface. In contrast to the above systems, *Axiom* has a fully typed language, and computations take place in specific mathematical contexts. MuPAD is the youngest among the general purpose systems. It has been

---

[1] MuPAD's programming language is structured similarly to Pascal. There exists a concept for object oriented programming.

[2] MuPAD uses parts of this package internally.

developed at the University of Paderborn since 1990, and it tries to combine the merits of its predecessors with modern concepts of its own.

1.5 MuPAD

In addition to the stated properties of computer algebra systems, Mu-PAD has the following features:

- MuPAD provides a concept for object oriented programming. You can define your own data types. Almost all existing operators and functions can be overloaded.
- MuPAD offers an interactive source code debugger.
- Programs written in C or C++ can be added to the kernel by Mu-PAD's dynamic module concept.

The heart of MuPAD is its *kernel*, which is implemented in C and partly in C++. It comprises the following main components:

- The *parser* reads the input to the system and performs a syntax check. If no errors are found, it converts the input to a MuPAD data type.
- The *evaluator* processes and simplifies the input. Its mode of operation is discussed later.
- The *memory management* is responsible for the efficient storage of MuPAD objects.
- Some frequently used algorithms such as, e.g., arithmetical functions are implemented as kernel functions in C.

Moreover, MuPAD's programming language is defined within the kernel. The MuPAD libraries, which contain most of the mathematical knowledge of the system, are implemented in this language.

In addition, MuPAD offers comfortable user interfaces for generating *notebooks* or graphics, or to debug programs written in MuPAD's language. The MuPAD help system has hypertext functionality. You can navigate within documents and execute examples by a mouse click. Figure 1.1 shows the main components of the MuPAD system.

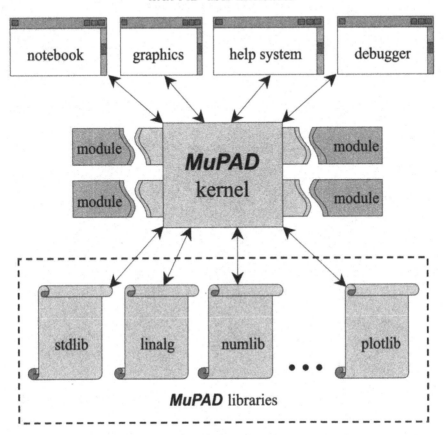

Figure 1.1. MuPAD's main components

2. First Steps in **MuPAD**

Computer algebra systems such as MuPAD are often used interactively. For example, you can enter an instruction to multiply two numbers together and wait until MuPAD computes the result and prints it on the screen.

After you call the MuPAD program, a *session* is launched. You find the information how to start the MuPAD program in the MuPAD installation instructions. MuPAD provides a help system which you can consult during a session to find details about system functions, their syntax, their parameters, etc. The following section presents an introduction to MuPAD's help system. Requesting a help page is probably the most frequently used command for the beginner. The next section is about using MuPAD as an "intelligent pocket calculator": calculating with numbers. This is the easiest and the most intuitive part of this tutorial. Afterwards we introduce some system functions for symbolic computations. The corresponding section is written quite informally and gives a first insight into the symbolic features of the system.

After starting the program, you can enter commands in the MuPAD language. The system awaits your input when the MuPAD prompt appears on the screen. On a Windows or Macintosh system, the prompt is the • sign, while it is >> on UNIX platforms. We use the UNIX prompt in all examples throughout the book. If you press the <RETURN> key under UNIX, this finishes your input and MuPAD evaluates the command that you have entered. On a Windows system, you need to press the combination of the <SHIFT> and the <RETURN> key to execute a command; pressing <RETURN> only provokes a linefeed and still leaves MuPAD in input mode[1]. On a Macintosh, <SHIFT>+<RETURN> or <ENTER> executes a command and <RETURN> only provokes a linefeed. Until MuPAD version 1.4, every command must be terminated by

---

[1] You can exchange the roles of <RETURN> and <SHIFT>+<RETURN> by choosing the "Options" item in the "View" menu and then clicking on "Enter only". In MuPAD versions beyond 1.4, "Enter only" is the default.

a semicolon or a colon before you can send it to the kernel by pressing <SHIFT>+<RETURN> (<RETURN> or <ENTER> on a UNIX or Macintosh system, respectively). Such a terminator is redundant in later versions.

If you enter:

```
>> sin(3.141);
```

and then press <SHIFT>+<RETURN> (<RETURN> or <ENTER>, respectively), the result:

```
   0.0005926535551
```

is printed on your screen. The system evaluates the usual sine function at the point 3.141 and returns a floating-point approximation of the value, similar to the output of a pocket calculator.

If you terminate your command with a colon instead of a semi-colon, MuPAD executes the command without printing its result on the screen. This enables you to suppress the output of irrelevant intermediate results. You can enter more than one command in a row. Two subsequent commands have to be separated by a semicolon or a colon:

```
>> diff(sin(x^2), x); int(last(1), x);

           2
   2 x cos(x )


         2
   sin(x )
```

Here x^2 denotes the square of x, and the MuPAD functions diff and int perform the operations "differentiate" and "integrate" (Chapter 7). The command last(1) returns the previous expression (in the example, this is the derivative of $\sin(x^2)$). The concept underlying last is discussed in Chapter 12.

In the following example, the output of the first command is suppressed by the semicolon, and only the result of the second command appears on the screen:

```
>> equations := {x + y = 1, x - y = 1}:
>> solve(equations);
```

```
{{x = 1, y = 0}}
```

In the previous example, a set of two equations is assigned to the identifier equations. The command solve(equations) computes the solution. Chapter 8 discusses the solver in more detail.

You can end the current MuPAD session by entering the keyword quit:

```
>> quit
```

On a Windows system, this terminates only the MuPAD kernel, and you can quit the user interface from the menu of the MuPAD window.

2.1 Explanations and Help

If you do not know the correct syntax of a MuPAD command, then you can obtain this information directly from the online help system. For many MuPAD routines, the function info returns a brief explanation:

```
>> info(solve);
```

```
solve -- solve equations and inequations [try ?solve \
for options]
```

```
>> info(ln);
```

```
ln -- the natural logarithm
```

The *help page* of the corresponding function provides more detailed information. You can request it by entering help("name"), where name is the name of the function. The function help expects its argument to be a string, which are generated by double quotes " in MuPAD (Section 4.11). The operator ? is a short form for help. It can be used without parenthesis or quotes:

```
>> ?solve
```

In version 1.4, requesting help by ? is one of the few exceptional commands that need not be terminated by a semicolon or a colon[2]. The layout of the help pages depends on the MuPAD version. In the following example, you can see a help page in ASCII format, like it is returned by the terminal version of MuPAD in response to ?solve:

```
solve -- general solver

Calling Sequence:

solve( equ , indet )
solve( sys , unk <MaxDegree=n> <BackSubstitution=b> )
solve( object )

Parameters:

equ -- equation or expression
indet -- indeterminate
sys -- set of equations
unk -- set of indeterminates
n -- nonnegative integer
b -- TRUE or FALSE
object -- domain element

Overloadable:

equ , sys , object

Description:

solve( equ, indet ) solves the equation equ with
respect to the variable indet and returns a set
of solutions. ...
```

We omit the remainder of the output to save space. Figure 2.1 shows a part of the corresponding hypertext document that appears if you have a graphical user interface. By clicking with the mouse in this help window, you can jump to the list of available functions, to the index,

---

[2] Further exceptions are the commands quit, which exits MuPAD, and !, which is used to call the operating system from within a MuPAD session; see Section 14.4.

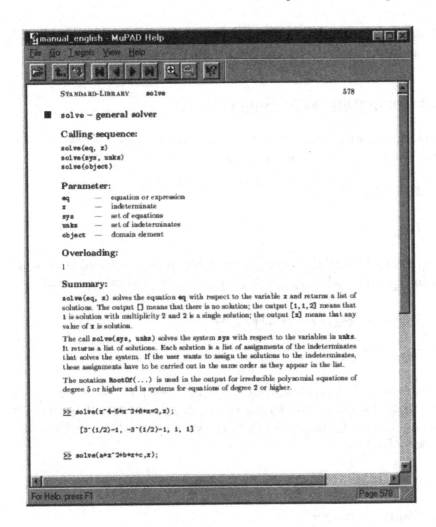

Figure 2.1. The help window in MuPAD Pro

or to the manual's table of contents. The help system is a hypertext system. Active keywords are underlined or framed. If you click on them, you obtain further information about the corresponding notion. Every page has some general links for requesting the previous or the next page, the beginning of the next chapter, or the index. On a Windows or Macintosh platform, you find these possibilities for navigation in the targets menu or the help menu, respectively. The examples in the help pages can be transferred to MuPAD's input window by clicking on the corresponding underlined or framed prompts.

Exercise 2.1: Find out how to use MuPAD's differentiator `diff`, and compute the fifth derivative of $\sin(x^2)$.

2.2 Computing with Numbers

To compute with numbers, you can use MuPAD like a pocket calculator. The result of the following input is a rational number:

```
>> 1 + 5/2;
```

   7/2

You see that MuPAD returns exact results (and not rounded floating-point numbers) when computing with integers and rational numbers:

```
>> (1 + (5/2*3))/(1/7 + 7/9)^2;
```

   67473/6728

The symbol `^` represents exponentiation. MuPAD can compute big numbers efficiently. The length of a number that you may compute is only limited by the available main storage. For example, the 123th power of 1234

```
>> 1234^123;
```

is a fairly big integer[3]:

```
17051580621272704287505972762062628265430231311106829\
04705296193221839138348680074713663067170605985726415\
92314554345900570589670671499709086102539904846514793\
13561730556366999395010462203568202735575775507008323\
84441477783960263870670426857004040032870424806396806\
96865587865016699383883388831980459159942845372414601\
80942971772610762859524340680101441852976627983806720\
3562799104
```

---

[3] In this printout, the "backslash" \ at the end of a row indicates that the result is continued on the next row.

Besides the basic arithmetic functions, MuPAD provides a variety of functions operating on numbers. A simple example is the factorial $n! = 1 \cdot 2 \cdot \ldots \cdot n$ of a nonnegative integer, which can be entered in mathematical notation:

```
>> 100!;
```

```
933262154439441526816992388562667004907159682643816 21\
468592963895217599993229915608941463976156518286253 69\
79208272237582511852109168640000000000000000000000 00
```

The function isprime checks whether a positive integer is prime. It returns either TRUE or FALSE. Using ifactor, you can obtain the prime factorization:

```
>> isprime(123456789);
```

```
FALSE
```

```
>> ifactor(123456789);
```

```
[1, 3, 2, 3607, 1, 3803, 1]
```

A look at the help page tells you how to interpret the returned list[4]: ?ifactor. We have $123456789 = +1 \cdot 3^2 \cdot 3607^1 \cdot 3803^1$. You obtain a more readable form of output by calling Factor instead. This function factors both symbolic expressions and integers:

```
>> Factor(123456789);
```

```
      2
   3   3607 3803
```

Now suppose that we want to "compute" the number $\sqrt{56}$. The problem is that the value of this irrational number cannot be expressed as a quotient numerator/denominator of two integers exactly. Thus "computation" can only mean to find an exact representation that is *as simple as possible*.

---

[4] In MuPAD versions beyond 1.4, the function ifactor returns a more readable output format (see the footnote on page 39), and the function Factor is obsolete.

2.2.1 Exact Computations

We try to illustrate the above problem of exact representation with some examples. When you input $\sqrt{56}$ via sqrt, then MuPAD returns the following:

```
>> sqrt(56);
```

$$2 \; 14^{1/2}$$

The result of the simplification of $\sqrt{56}$ is the exact value $2 \cdot 14^{1/2}$. Here, $14^{1/2}$ represents in MuPAD the positive solution of the equation $x^2 = 14$. Indeed, this is probably the most simple representation of the result. We stress that $14^{1/2}$ is a genuine MuPAD object with certain properties (namely that its square can be simplified to 14). The system applies them automatically when computing with such objects. For example:

```
>> sqrt(14)^4;
```

```
196
```

As another example for exact computation, let us determine the limit

$$e = \lim_{n \to \infty} \left(1 + \frac{1}{n}\right)^n.$$

We use the function limit and the identifier infinity:

```
>> limit((1 + 1/n)^n, n = infinity);
```

```
exp(1)
```

The identifier exp represents the exponential function. The Euler number e is represented exactly by the symbol[5] exp(1), and MuPAD knows the exact rules of manipulation for this object. For example, using the natural logarithm ln we find:

---

[5] The identifier E denotes the same object in MuPAD, namely the base of the natural logarithm ($e = \exp(1) = 2.71828..$).

```
>> ln(1/exp(1));
```

```
  -1
```

We will encounter more exact computations later in this tutorial.

2.2.2 Numerical Approximations

Besides exact computations, MuPAD can also perform numerical approximations. For example, you can use the MuPAD function float to find a decimal approximation to $\sqrt{56}$. This function computes the value of its argument in *floating-point representation*:

```
>> float(sqrt(56));
```

```
  7.483314773
```

The precision of the approximation depends on the value of the global variable DIGITS, which determines the number of decimal digits for numerical computations. Its default value is 10:

```
>> DIGITS; float(67473/6728);
```

```
  10
```

```
  10.02868608
```

Global variables such as DIGITS affect the behavior of MuPAD and are also called *environment variables*[6]. You find a complete list of all environment variables in Section "Environment Variables" of the MuPAD Quick Reference [Oev 98]. The variable DIGITS can assume any integral value between 1 and $2^{32} - 1$:

```
>> DIGITS := 100: float(67473/6728); DIGITS := 10:
```

---

[6] You should be particularly cautious when the same computation is performed with different values of DIGITS. Some of the more intricate numerical algorithms in MuPAD employ the option "remember". This implies that they store previously computed values to be used again (Section 18.8), which can lead to inaccurate numerical results if the remembered values were computed with lower precision. To be safe, you should restart the MuPAD session using reset() before increasing the value of DIGITS. This command erases MuPAD's memory and resets all environment variables to their default values (Section 14.3).

10.0286860879904875148632580261593341260404280618 3115\
33888228299643281807372175980975029726516 0523186

We have reset the value of DIGITS to 10 for the following computations. This can also be achieved via the command unassign(DIGITS)[7].
For arithmetic operations with numbers, MuPAD automatically uses
approximate computation as soon as *at least one* of the numbers involved is a floating-point value:

```
>> (1.0 + (5/2*3))/(1/7 + 7/9)^2;
```

```
   10.02868608
```

Please note that none of the two following calls

```
>> 2/3*sin(2), 0.6666666666*sin(2);
```

results in an approximate computation of $\sin(2)$, since technically
$\sin(2)$ is an expression representing the (exact) value of $\sin(2)$ and
not a number:

```
   2 sin(2)
   --------, 0.6666666666 sin(2)
      3
```

The separation of both values by a comma generates a special data
type, namely a *sequence*, which is described in Section 4.5. You have
to use the function **float** to compute a floating-point representation
of the above expressions[8]:

```
>> float(2/3*sin(2)), 0.6666666666*float(sin(2));
```

```
   0.6061982845, 0.6061982844
```

---

[7] The keyword **delete** replaces the function **unassign** in MuPAD versions beyond
1.4.

[8] Take a look at the last digits. The second command yields a slightly less accurate
result, since 0.666... is already an approximation of 2/3 and the rounding error
is propagated to the final result.

Most arithmetic functions in MuPAD, such as sqrt, the trigonometric functions, the exponential function, or the logarithm, automatically return approximate values when their argument is a floating-point number:

```
>> sqrt(56.0), sin(3.14);

   7.483314773, 0.001592652916
```

The constants π and e are denoted by PI and E = exp(1), respectively. MuPAD can perform exact computations with them:

```
>> cos(PI), ln(E);

   -1, 1
```

If desired, you can obtain numerical approximations of these constants by applying float:

```
>> DIGITS := 100: float(PI); float(E); unassign(DIGITS):

   3.141592653589793238462643383279502884197169399375105\
   8209749445923078164062862089986280348253421170680

   2.718281828459045235360287471352662497757247093699959\
   5749669676277240766303535475945713821785251664270
```

Exercise 2.2: Compute $\sqrt{27} - 2\sqrt{3}$ and $\cos(\pi/8)$ exactly. Determine numerical approximations to a precision of 5 digits.

2.2.3 Complex Numbers

The imaginary unit $\sqrt{-1}$ is represented in MuPAD by the symbol I:

```
>> sqrt(-1), I^2;

   I, -1
```

You can input complex numbers in MuPAD in the usual mathematical notation $x + y\,I$. Both the real part x and the imaginary part y may be integers, rational numbers, or floating-point numbers:

```
>> (1 + 2*I)*(4 + I), (1/2 + I)*(0.1 + I/2)^3;

   2 + 9 I, 0.073 - 0.129 I
```

If you use symbolic expressions such as, e.g., sqrt(2), then MuPAD may not return the result of a calculation in Cartesian coordinates:

```
>> 1/(sqrt(2) + I);

      1
   --------
    1/2
   2    + I
```

The function rectform (short for: rectangular form) ensures that the result is split into its real and imaginary parts:

```
>> rectform(1/(sqrt(2) + I));

    1/2
   2
   ---- + (-1/3) I
    3
```

The functions Re and Im return the real part x and the imaginary part y, respectively, of a complex number $x + y\,I$. The MuPAD functions conjugate and abs compute the complex conjugate $x - y\,I$ and the absolute value $|x + y\,I| = \sqrt{x^2 + y^2}$, respectively:

```
>> Re(1/(sqrt(2) + I)), Im(1/(sqrt(2) + I)),
   abs(1/(sqrt(2) + I)), conjugate(1/(sqrt(2) + I)),
   rectform(conjugate(1/(sqrt(2) + I)));

    1/2            1/2                1/2
   2              3           1      2
   ----, -1/3, ----, --------, ---- + 1/3 I
    3            3      1/2      3
                       2    - I
```

2.3 Symbolic Computation

This section comprises some examples of MuPAD sessions that illustrate a small selection of the system's power of symbolic manipulation. The mathematical knowledge is contained essentially in MuPAD's functions for differentiation, integration, simplification of expressions, etc. This demonstration does not proceed in a particularly systematic manner: we apply the system functions to objects of various types, such as sequences, sets, lists, expressions etc. Those are explained in Chapter 4 one by one and in detail.

2.3.1 Introductory Examples

A symbolic expression in MuPAD may contain undetermined quantities (identifiers). The following expression contains two unknowns x and y:

```
>> f := y^2 + 4*x + 6*x^2 + 4*x^3 + x^4;
```

$$4\ x + 6\ x^2 + 4\ x^3 + y^2 + x^4$$

Using the assignment operator :=, we have assigned the expression to an identifier f, which can now be used as an abbreviation for the expression. We say that the latter is the *value* of the identifier f. We note that MuPAD has exchanged the order of the terms[9].

MuPAD offers the system function diff for differentiating expressions:

```
>> diff(f, x), diff(f, y);
```

$$12\ x + 12\ x^2 + 4\ x^3 + 4,\ 2\ y$$

Here, we have computed both the derivative with respect to x and to y. You may obtain higher derivatives either by nested calls of diff, or by a single call:

---

[9] Internally, symbolic sums are ordered according to certain rules that enable the system to access the terms faster. Of course, such a reordering of the input happens only for commutative operations such as, e.g., addition or multiplication, where changing the order of the operands yields a mathematically equivalent object.

```
>> diff(diff(diff(f, x), x), x), diff(f, x, x, x);
```

```
24 x + 24, 24 x + 24
```

Alternatively, you can use the differential operator ', which maps a function to its derivative[10]:

```
>> sin', sin'(x);
```

```
cos, cos(x)
```

The symbol ' for the derivative is a short form of the differential operator D. The call D(function) returns the derivative:

```
>> D(sin), D(sin)(x);
```

```
cos, cos(x)
```

You can compute integrals by using int. The following command computes a definite integral on the real interval between 0 and 1:

```
>> int(f, x = 0..1);
```

```
 2
y  + 26/5
```

The next command determines an indefinite integral and returns an expression containing the integration variable x and a symbolic parameter y:

---

[10] MuPAD uses a mathematically strict notation for the differential operator: D differentiates functions, while diff differentiates expressions. In the example, D maps the (name of the) function to the (name of the) function representing the derivative. You often find a sloppy notation such as, e.g., $(x + x^2)'$ for the derivative of the function $F : x \mapsto x + x^2$. This notation confuses the map F and the image point $f = F(x)$ at a point x. MuPAD has a strict distinction between the *function* F and the *expression* $f = F(x)$, which are realized as different data types. The map corresponding to f can be defined by

```
>> F := x -> (x + x^2):
```

Then

```
>> diff(f,x) = F'(x);
```

```
2 x + 1 = 2 x + 1
```

are equivalent ways of obtaining the derivative as expressions. The call f:=x+x^2; f'; does not make sense in MuPAD.

```
>> int(f, x);
```

$$2x^2 + 2x^3 + x^4 + \frac{x^5}{5} + x^2 y$$

If you try to compute the indefinite integral of an expression and it cannot be represented by elementary functions, then `int` returns the call unevaluated:

```
>> integral := int(1/(exp(x^2) + 1), x);
```

```
        /      1         \
   int |  ----------- , x |
        |       2          |
        \ exp(x ) + 1     /
```

Nevertheless, this object has mathematical properties. The differentiator recognizes that its derivative is the integrand:

```
>> diff(integral, x);
```

```
       1
   -----------
        2
   exp(x ) + 1
```

Definite integrals may also be returned unevaluated by `int`:

```
>> int(1/(exp(x^2) + 1), x = 0..1);
```

```
        /      1                \
   int |  ----------- , x = 0..1 |
        |       2                 |
        \ exp(x ) + 1            /
```

The corresponding mathematical object is a real number, and the output is an exact representation of this number which MuPAD was unable to simplify further. As usual, you can obtain a floating-point approximation by applying `float`:

```
>> float(%);
```

```
    0.41946648
```

The symbol % (which is equivalent to last(1)) is an abbreviation for the previously computed expression (Chapter 12).

MuPAD knows the most important mathematical functions such as the square root sqrt, the exponential function exp, the trigonometric functions sin, cos, tan, the hyperbolic functions sinh, cosh, tanh, the corresponding inverse functions ln, asin, acos, atan, asinh, acosh, atanh[11], as well as a variety of other special functions such as, e.g., the gamma function, the error function erf, Bessel functions, etc. (Section "Special Mathematical Functions" of the MuPAD Quick Reference [Oev 98] gives a survey.) In particular, MuPAD knows the rules of manipulation for these functions (e.g., the addition theorems for the trigonometric functions) and applies them. It can compute floating-point approximations such as, e.g., float(exp(1))=2.718.., and knows special values (e.g., sin(PI)=0). If you call these functions, they often return themselves unevaluated, since this is the most simple exact representation of the corresponding value:

```
>> sqrt(2), exp(1), sin(x + y);
```

```
    1/2
    2   , exp(1), sin(x + y)
```

The main feature of the system is to simplify or transform such expressions using the rules for computation. For example, the system function **expand** "expands" functions such as exp, sin, etc. by means of the addition theorems if their argument is a symbolic sum:

```
>> expand(exp(x + y)), expand(sin(x + y)),
   expand(tan(x + 3*PI/2));
```

```
                                            cos(x)
   exp(x) exp(y), cos(x) sin(y) + cos(y) sin(x), - ------
                                            sin(x)
```

---

[11] In MuPAD versions beyond 1.4, these inverse functions are renamed to arcsin, arccos, arctan, arcsinh, arccosh, arctanh.

Generally speaking, one of the main tasks of a computer algebra system is to manipulate and to simplify expressions. Besides `expand`, MuPAD provides the functions `collect`, `combine`, `normal`, `partfrac`, `radsimp`, `rewrite`, and `simplify` for manipulation. They are presented in greater detail in Chapter 9. We briefly mention some of them in what follows.

The function `normal` finds a common denominator for rational expressions:

```
>> f := x/(1 + x) - 2/(1 - x): g := normal(f);
```

$$\frac{x + x^2 + 2}{x^2 - 1}$$

Moreover, `normal` automatically cancels common factors in numerator and denominator:

```
>> normal(x^2/(x + y) - y^2/(x + y));
```

$$x - y$$

Conversely, `partfrac` (short for "partial fraction") decomposes a rational expression into a sum of rational terms with simple denominators:

```
>> partfrac(g, x);
```

$$\frac{2}{x - 1} - \frac{1}{x + 1} + 1$$

The function `simplify` is a universal simplifier and tries to find a representation that is as simple as possible:

```
>> simplify((exp(x) - 1)/(exp(x/2) + 1));
```

```
   / x \
expl - | - 1
   \ 2 /
```

You may control the simplification by supplying `simplify` with additional arguments (see `?simplify`). The function `radsimp` simplifies arithmetical expressions containing radicals (roots):

```
>> f := sqrt(4 + 2*sqrt(3)): f = radsimp(f);
```

```
  1/2   1/2      1/2      1/2
 2    (3    + 2)     = 3     + 1
```

Here, we have generated an equation, which is a genuine MuPAD object. Another important function is `Factor`[12], which decomposes an expression into a product of simpler ones:

```
>> Factor(x^3 + 3*x^2 + 3*x + 1),
   Factor(2*x*y - 2*x - 2*y + x^2 + y^2),
   Factor(x^2/(x + y) - z^2/(x + y));
```

```
         3                        (x + z) (x - z)
  (x + 1) , (x + y) (x + y - 2), ----------------
                                      x + y
```

The function `limit` does what its name suggests. For example, the function $\sin(x)/x$ has a removable pole at $x = 0$. Its limit for $x \to 0$ is 1:

```
>> limit(sin(x)/x, x = 0);
```

```
1
```

In a MuPAD session, you can define functions of your own in several ways. A simple and intuitive method is to use the arrow operator `->` (the minus symbol followed by the "greater than" symbol):

---

[12] The function `Factor` is replaced by `factor` in MuPAD versions beyond 1.4; see the footnote on page 135.

```
>> F := x -> (x^2):  F(x), F(y), F(a + b), F'(x);
```

$$x^2, \ y^2, \ (a + b)^2, \ 2x$$

In Chapter 18, we discuss MuPAD's programming features and describe how to implement more complex algorithms as MuPAD procedures.

In window-based MuPAD versions, you can use the graphics facilities to visualize mathematical objects immediately. The relevant MuPAD functions for generating graphics are plotfunc[13], plot2d, plot3d, and the routines from the graphics library plotlib. You can let MuPAD draw the graphs of functions with one or two arguments by using plotfunc:

```
>> plotfunc(sin(x^2), x = -2..5);
```

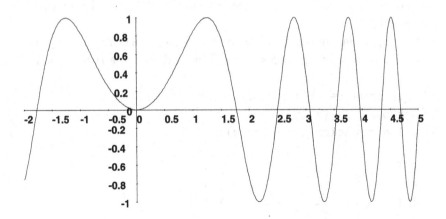

```
>> plotfunc(sin(x^2 + y^2), x = 0..PI, y = 0..PI);
```

---

[13] The function plotfunc is replaced by plotfunc2d and plotfunc3d in MuPAD versions beyond 1.4.

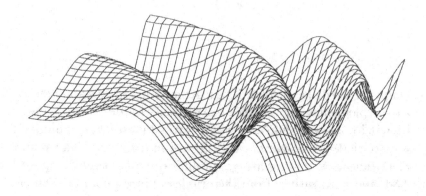

Depending on your MuPAD version, either the graphics module opens a separate window, or the plot appears in the notebook below the call of the graphics command. You can manipulate the graphics interactively. Alternatively, you may supply **plot2d** or **plot3d** directly with the desired arguments. You find a description of the graphics features in Chapter 11.

Solving equations or systems of equations is certainly an important task for a computer algebra system. This is done via **solve** in MuPAD:

```
>> equations := {x + y = a, x - a*y = b}:
>> unknowns := {x, y}:
>> solve(equations, unknowns);
```

```
  { {            2                  } }
  { {        b + a         a - b } }
  { { x = ------, y = ----- } }
  { {        a + 1         a + 1 } }
```

Here, we have generated a set of two equations and a set of unknowns which we wish to solve for. MuPAD returns the result in terms of simplified equations, from which you can read off the solution. In the above example, there are two more symbolic parameters a and b. This is why we have told **solve** which of the symbols it should express in terms of the others. In the following example, we have only one equation in one unknown. MuPAD automatically recognizes the unknown and solves for it:

```
>> solve(x^2 - 2*x + 2 = 0, x);
```

```
{1 - I, 1 + I}
```

The result is a set containing the two (complex) solutions of the quadratic equation. You find a detailed description of solve in Chapter 8.

The functions sum and product handle symbolic sums and products. For example, the well-known sum $1 + 2 + \cdots + n$ is:

```
>> sum(i, i = 1..n); Factor(%);
```

$$\frac{n^2}{2} + \frac{n}{2}$$

$$\frac{n\,(n + 1)}{2}$$

The product $1 \cdot 2 \cdot \ldots \cdot n$ is better known as factorial $n!$ and can be computed via MuPAD's fact function. More generally, MuPAD uses the gamma function, which is defined for non-integral or even complex arguments. If n is a nonnegative integer, then we have gamma(n+1) = $n! = $ fact(n):

```
>> product(i^3, i = 1..n); rewrite(%, fact);
```

$$gamma(n + 1)^3$$

$$fact(n)^3$$

The command rewrite(expression,fact) replaces all symbolic occurrences of the gamma function by symbolic calls of the factorial fact (Section 9.1).

There exist several data structures for vectors and matrices in MuPAD. In principle, you may use arrays (Section 4.9) to represent

such objects, but it is far more intuitive to work with the data type
"matrix". You can generate matrices by using the system function[14]
Dom::Matrix(). For easier use, we introduce a suggestive abbreviation:

```
>> constructor := Dom::Matrix():
```

Now you can generate a matrix as follows:

```
>> A := constructor([[1, 2], [a, 4]]);
```

```
+-      -+
| 1, 2 |
|       |
| a, 4 |
+-      -+
```

So constructed objects have the convenient property that the basic
arithmetic operations +, *, etc. are redefined ("overloaded") according
to the appropriate mathematical context. For example, you may use
+ or * to add or multiply matrices, respectively (if the dimensions
match):

```
>> B := constructor([[y, 3], [z, 5]]):
>> A, B, A + B, A*B;
```

```
+-      -+ +-      -+ +-          -+
| 1, 2 | | y, 3 | | y + 1, 5 |
|      |,| |      |,| |          |,
| a, 4 | | z, 5 | | a + z, 9 |
+-      -+ +-      -+ +-          -+
```

```
+-                    -+
|   y + 2 z,      13   |
|                      |
| 4 z + a y, 3 a + 20 |
+-                    -+
```

---

[14] Dom::Matrix is itself a function, which may receive the type of the matrix entries as an optional argument. In this way, you can define matrices whose entries must be integers, rational numbers, floating-point numbers, polynomials etc.
(Section 4.15). If the optional argument is missing, as in this example, then the
generated matrices may have essentially arbitrary MuPAD expressions as entries.

The power `A^(-1)` denotes the inverse of the matrix `A`:

```
>> A^(-1);
```

```
+-                                    -+
|       2 a                   2        |
|    ----------  + 1, -   ----------   |
|    - 2 a + 4            - 2 a + 4    |
|                                      |
|         a                   1        |
|    - ----------,        ----------   |
|      - 2 a + 4          - 2 a + 4    |
+-                                    -+
```

The function `linalg::det`, from MuPAD's `linalg` library for linear algebra (Section 4.15.4), computes the determinant:

```
>> linalg::det(A);
```

```
   4 - 2 a
```

Column vectors of dimension n can be interpreted as $n \times 1$ matrices:

```
>> b := constructor([1, x]);
```

```
+-    -+
|  1  |
|     |
|  x  |
+-    -+
```

You can comfortably determine the solution $A^{-1}\mathbf{b}$ of the system of linear equations $A\mathbf{x} = \mathbf{b}$, with the above coefficient matrix `A` and the previously defined `b` on the right hand side:

```
>> solutionVector := A^(-1)*b;
```

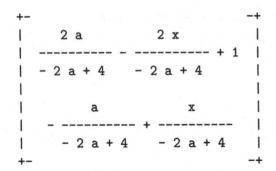

Now you can apply the function **normal** to each component of the vector by means of the system function **map**, thus simplifying the representation:

```
>> map(%, normal);
```

```
+-          -+
|   x - 2   |
|   -----   |
|   a - 2   |
|           |
|   a - x   |
|  -------  |
|  2 a - 4  |
+-          -+
```

To verify MuPAD's computation, you may multiply the matrix **A** by the solution vector:

```
>> A*%;
```

```
+-                        -+
|    x - 2    2 (a - x)    |
|    -----  + ---------    |
|    a - 2    2 a - 4      |
|                          |
|  a (x - 2)    4 (a - x)  |
|  ---------  + ---------  |
|    a - 2      2 a - 4    |
+-                        -+
```

After a simplification, you can check that the result equals b:

```
>> map(%, normal);
```

```
+-    -+
|  1  |
|     |
|  x  |
+-    -+
```

Section 4.15 provides more information on handling matrices and vectors.

Exercise 2.3: Compute an expanded form of the expression $(x^2 + y)^5$.

Exercise 2.4: Use MuPAD to check that $\dfrac{x^2 - 1}{x + 1} = x - 1$ holds.

Exercise 2.5: Generate a plot of the function $f(x) = 1/\sin(x)$ for $1 \le x \le 10$.

Exercise 2.6: Obtain detailed information about the function `limit`. Use MuPAD to verify the following limits:

$$\lim_{x \to 0} \frac{\sin(x)}{x} = 1 \ , \quad \lim_{x \to 0} \frac{1 - \cos(x)}{x} = 0 \ , \quad \lim_{x \to 0+} \ln(x) = -\infty \ ,$$

$$\lim_{x \to 0} x^{\sin(x)} = 1 \ , \quad \lim_{x \to \infty} \left(1 + \frac{1}{x}\right)^x = e \ , \quad \lim_{x \to \infty} \frac{\ln(x)}{e^x} = 0 \ ,$$

$$\lim_{x \to 0} x^{\ln(x)} = \infty \ , \quad \lim_{x \to 0} \left(1 + \frac{\pi}{x}\right)^x = e^\pi \ , \quad \lim_{x \to 0-} \frac{2}{1 + e^{-1/x}} = 0 \ .$$

The limit $\lim\limits_{x \to 0} \sin(x)^{1/x}$ does not exist. How does MuPAD react?

Exercise 2.7: Obtain detailed information about the function `sum`. The call `sum(f(k),k=a..b)` computes a *closed form* of a finite or infinite sum, if possible. Use MuPAD to verify the following identity:

$$\sum_{k=1}^{n} (k^2 + k + 1) = \frac{n\,(n^2 + 3\,n + 5)}{3} \ .$$

Determine the values of the following series:

$$\sum_{k=0}^{\infty} \frac{2\,k - 3}{(k + 1)\,(k + 2)\,(k + 3)} \ , \quad \sum_{k=2}^{\infty} \frac{k}{(k - 1)^2\,(k + 1)^2} \ .$$

Exercise 2.8: Compute $2 \cdot (A + B)$, $A \cdot B$, and $(A - B)^{-1}$ for the following matrices:

$$A = \begin{pmatrix} 1 & 2 & 3 \\ 4 & 5 & 6 \\ 7 & 8 & 0 \end{pmatrix}, \quad B = \begin{pmatrix} 1 & 1 & 0 \\ 0 & 0 & 1 \\ 0 & 1 & 0 \end{pmatrix}.$$

2.3.2 Curve Sketching

In the following sample session, we use some of the system functions from the previous section to sketch and discuss the curve given by the rational function:

$$f : x \mapsto \frac{(x - 1)^2}{x - 2} + a$$

with a parameter a. At first we determine some characteristics of this function.

```
>> f := x -> ((x - 1)^2/(x - 2) + a):
>> singularities := discont(f(x), x);
```

    {2}

The function `discont` determines the discontinuities of the function f with respect to the variable x. It returns a set of such points. Thus the above f is defined and continuous for all $x \neq 2$. Obviously $x = 2$ is a pole. Indeed, MuPAD finds the limit $\pm\infty$ when you approach this point from the left or from the right, respectively:

```
>> limit(f(x), x = 2, Left), limit(f(x), x = 2, Right);
```

    -infinity, infinity

You find the roots of f by solving the equation $f = 0$:

```
>> roots := solve(f(x) = 0, x);
```

```
{                    2 1/2              2 1/2            }
{      a    (4 a + a )          (4 a + a )       a      }
{ 1 -  - -  --------------, --------------  - - + 1 }
{      2        2                2              2       }
```

Depending on a, either both or none of the two roots are real. Now we want to find the local extrema of f. To this end, we determine the roots of the first derivative f':

```
>> f'(x);
```

```
                2
  2 x - 2   (x - 1)
  ------- - --------
   x - 2        2
            (x - 2)
```

```
>> extrema := solve(f'(x) = 0, x);
```

```
  {1, 3}
```

These are the candidates for local extrema. However, some of them might be saddle points. If the second derivative f'' does not vanish at these points, then both are really extrema. We check:

```
>> f''(1), f''(3);
```

```
  -2, 2
```

Our results imply that f has the following properties: for any choice of the parameter a, there is a local maximum at $x = 1$, a pole at $x = 2$, and a local minimum at $x = 3$. The corresponding values of f at these points are

```
>> maxvalue := f(1); minvalue := f(3);
```

```
  a
```

```
  a + 4
```

f tends to $\mp\infty$ for $x \to \mp\infty$:

```
>> limit(f(x), x = -infinity), limit(f(x), x = infinity);
```

```
  -infinity, infinity
```

We can specify the behavior of f more precisely for large values of x.
It asymptotically approaches the linear function $x \mapsto x + a$:

```
>> series(f(x), x = infinity);
```

```
            1   2    4      / 1 \
    x + a + - + -- + -- + 0|  -- |
            x    2    3     |  4  |
                x    x      \ x  /
```

Here we have employed the function `series` to compute an asymptotic
expansion of f (Section 4.13). We can easily check our results visually
by plotting the graph of f for several values of a:

```
>> F := subs(f(x), a = -4): G := subs(f(x), a = 0):
   H := subs(f(x), a = 4): F, G, H;
```

```
         2              2          2
    (x - 1)         (x - 1)    (x - 1)
    --------- - 4,  ---------,  --------- + 4
     x - 2           x - 2      x - 2
```

The function `subs` (Chapter 6) replaces subexpressions: in the example, we have substituted the concrete values $-4, 0$, and 4, respectively,
for a. We now can plot the three functions together in one picture:

```
>> plotfunc(F, G, H, x = -1..4);
```

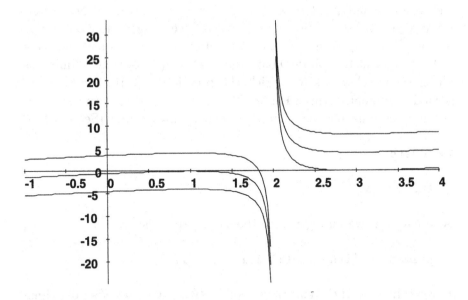

2.3.3 Elementary Number Theory

MuPAD provides a lot of elementary number theoretic functions, for example:

- `isprime(n)` tests whether $n \in \mathbb{N}$ is a prime number,
- `ithprime(n)` returns the nth prime number,
- `nextprime(n)` finds the least prime number $\geq n$,
- `ifactor(n)` computes the prime factorization of n.

These routines are quite fast. However, since they employ probabilistic primality tests, they may return wrong results, with very small probability[15]. Instead of `isprime`, you can use the (slower) function `numlib::proveprime` as an error-free primality test.

Let us generate a list of all primes up to 10000. Here is one of the many ways to do this:

```
>> primes := select([$ 1..10000], isprime);
```

```
    [2, 3, 5, 7, 11, 13, 17, ... , 9949, 9967, 9973]
```

---

[15] In practice, you need not worry about this because the chances of a wrong answer are negligible: the probability of a hardware failure is much higher than the probability that the randomized test returns the wrong answer on a correctly working hardware.

The result is not printed completely for space considerations. First, we have generated the sequence of all positive integers up to 10000 by means of the sequence generator $ (Section 4.5). The square brackets [ ] convert this to a MuPAD list. Then **select** (Section 4.6) eliminates all those list elements for which the function **isprime**, supplied as second argument, returns **FALSE**. The number of these primes equals the number of list elements, which we can obtain via **nops** (Section 4.1):

```
>> nops(primes);
```

```
   1229
```

Alternatively, we may generate the same prime list by

```
>> primes := [ithprime(i) $ i = 1..1229]:
```

Here we have used the fact that we already know the number of primes up to 10000. Another possibility is to generate a large list of primes and discard the ones greater than 10000:

```
>> primes := select([ithprime(i) $ i=1..5000],
                x -> (x<=10000)):
```

Here, the object **x -> (x<=10000)** represents the function that maps each **x** to the inequality **x<=10000**. The **select** command then keeps only those list elements for which the inequality evaluates to **TRUE**.

In the next example, we use a **repeat** loop (Chapter 16) to generate the list of primes. With the help of the concatenation operator **.** (Section 4.6), we successively append primes i to a list until **nextprime(i+1)**, the next prime greater than i, exceeds 10000. We start with the empty list and the first prime $i = 2$:

```
>> primes := [ ]: i := 2:
>> repeat
       primes := primes . [i];
       i := nextprime(i + 1)
    until i > 10000 end_repeat:
```

Now we consider Goldbach's famous conjecture:

"Every even integer greater than 2 is the sum of two primes."

We want to verify this conjecture for all even numbers up to 10000. First, we generate the list of integers [4,6,..,10000]. To reuse the variable i, we delete its value from the previous **repeat** loop by calling unassign[16].

```
>> unassign(i): list := [2*i $ i = 2..5000]:
>> nops(list);
```

    4999

Now we select those numbers from the list that cannot be written in the form "prime + 2". This is done by testing for each i in the list whether $i - 2$ is a prime:

```
>> list := select(list, i -> (not isprime(i - 2))):
>> nops(list);
```

    4998

The only integer that has been eliminated is 4 (since for all other even positive integers $i-2$ is even and greater than 2, and hence not prime). Now we discard all numbers of the form "prime + 3":

```
>> list := select(list, i -> (not isprime(i - 3))):
>> nops(list);
```

    3770

The remaining 3770 integers are neither of the form "prime + 2" nor of the form "prime + 3". We now continue this procedure by means of a **while** loop (Chapter 16). In the loop, j successively runs through all primes > 3, and the numbers of the form "prime + j" are eliminated. A **print** command (Section 13.1.1) outputs the number of remaining integers in each step. The loop ends as soon as the list is empty:

---

[16] The variable to the right of the sequence operator $ must have no value in MuPAD version 1.4, otherwise an error is provoked. In later versions, bound variables may already have a value, which is ignored by $.

```
>> j := 3:
>> while list <> [] do
      j := nextprime(j + 1):
      list := select(list, i -> (not isprime(i - j))):
      print(j, nops(list)):
   end_while:
```

```
    5, 2747

    7, 1926

    ...

  167, 1

  173, 0
```

Thus we have confirmed that Goldbach's conjecture holds true for all even positive integers up to 10000. We have even shown that all those numbers can be written as a sum of a prime less or equal to 173 and another prime.

In the next example, we generate a list of distances between two successive primes up to 500:

```
>> primes := select([$ 1..500], isprime):
>> distances := [primes[i] - primes[i - 1]
                  $ i = 2..nops(primes)];
```

```
[1, 2, 2, 4, 2, 4, 2, 4, 6, 2, 6, 4, 2, 4, 6, 6, 2, 6,

   4, 2, 6, 4, 6, 8, 4, 2, 4, 2, 4, 14, 4, 6, 2, 10, 2

   , 6, 6, 4, 6, 6, 2, 10, 2, 4, 2, 12, 12, 4, 2, 4, 6

   , 2, 10, 6, 6, 6, 2, 6, 4, 2, 10, 14, 4, 2, 4, 14,

   6, 10, 2, 4, 6, 8, 6, 6, 4, 6, 8, 4, 8, 10, 2, 10,

   2, 6, 4, 6, 8, 4, 2, 4, 12, 8, 4, 8]
```

The indexed call `primes[i]` returns the ith element in the list. The function `zip` (Section 4.6) provides an alternative method. The call `zip(a,b,f)` combines two lists $a = [a_1, a_2, \ldots]$ and $b = [b_1, b_2, \ldots]$ componentwise by means of the function f: the resulting list is

$$[f(a_1, b_1), f(a_2, b_2), \ldots]$$

and has as many elements as the shortest of the two lists. In our example, we apply this to the prime list $a = [a_1, \ldots, a_n]$, the "shifted" prime list $b = [a_2, \ldots, a_n]$, and the function $(x, y) \mapsto y - x$. We first generate a shifted copy of the prime list by deleting the first element, thus shortening the list:

```
>> b := primes: unassign(b[1]):
```

The following command returns the same result as above:

```
>> distances := zip(primes, b, (x, y) -> (y - x)):
```

We have presented another useful function in Section 2.2, the routine `ifactor` for factoring an integer into primes. The call `ifactor(n)` returns a list[17]

$$[s, p_1, e_1, \ldots, p_k, e_k]$$

with primes $p_1, \ldots, p_k$, their exponents $e_1, \ldots, e_k$, and the sign $s = \pm 1$, such that $n = s \cdot p_1^{e_1} \cdot p_2^{e_2} \cdots p_k^{e_k}$. We now employ this function to find out how many integers between 2 and 10000 are divisible by exactly two distinct prime numbers. We note that the list returned by `ifactor(n)` has $2m + 1$ elements, where m is the number of distinct prime divisors of n. Thus the function

```
>> m := (nops@ifactor - 1)/2:
```

returns the number of distinct prime factors. The symbol `@` generates the composition (Section 4.12) of the two functions `ifactor` and `nops`. Thus the call `m(k)` returns `m(k)=(nops(ifactor(k))-1)/2`. We construct the list of values $m(k)$ for $k = 2, \ldots, 10000$:

---

[17] In MuPAD versions beyond 1.4, `ifactor` returns a MuPAD object of domain type `Factored`. It is printed on the screen in a more readable form. Internally, however, the prime factors and the exponents are still stored in form of a list, and you can access them in the same way as in MuPAD 1.4, by using `op` or an indexed access. Consult the help pages `?ifactor` and `?Factored` for details.

```
>> list := [m(k) $ k = 2..10000]:
```

The following **for** loop (Section 16) displays the number of integers with precisely $i = 1, 2, \ldots, 6$ distinct prime divisors:

```
>> for i from 1 to 6 do
       print(i, nops(select(list, x -> (x = i))))
   end_for:
```

1, 1280

2, 4097

3, 3695

4, 894

5, 33

6, 0

Thus there are 1280 integers with exactly one prime divisor in the scanned interval[18], 4097 integers with precisely two distinct prime factors, and so on. It is easy to see that the interval contains no integer with six prime divisors: the smallest such number $2 \cdot 3 \cdot 5 \cdot 7 \cdot 11 \cdot 13 = 30030$ exceeds 10000.

The **numlib** library comprises various number theoretic functions. Among others, it contains the routine **numlib::numprimedivisors**, equivalent to the above **m**, for computing the number of prime divisors. We refer to Chapter 3 for a description of the MuPAD libraries.

Exercise 2.9: Primes of the form $2^n \pm 1$ always have produced particular interest.

a) Primes of the form $2^p - 1$, where p is a prime, are called *Mersenne primes*. Find all Mersenne primes for $1 < p \le 1000$.

b) For a positive integer n, the *nth Fermat number* is $2^{(2^n)} + 1$. Refute Fermat's conjecture that all those numbers are primes.

---

[18] We have already seen that the interval contains 1229 prime numbers. Can you explain the difference?

3. The **MuPAD** Libraries

Most of MuPAD's mathematical knowledge is organized in libraries. Such a library comprises a collection of functions for solving problems in a particular area, such as linear algebra, number theory, numerical analysis, etc. Library functions are written in MuPAD's programming language. You can use them in the same way as kernel functions, without knowing anything about the programming language.

Section "Libraries" of the MuPAD Quick Reference [Oev 98] contains a survey of all libraries. They are developed continuously, and future MuPAD versions provide additional functionality. In this chapter we do not address the mathematical functionality of libraries but rather discuss their general use.

3.1 Information About a Particular Library

You can obtain information and help for libraries by calling the functions **info** and **help**. The function **info** gives a list of all functions installed in the library. The **numlib** library is a collection of number theoretic functions:

```
>> info(numlib);
```

```
    Library 'numlib':        the package for elementary
    number theory
    Interface:
    numlib::Lambda,          numlib::Omega,
    numlib::decimal,         numlib::divisors,
    numlib::ecm,             numlib::fibonacci,
    numlib::fromAscii,       numlib::g_adic,

      ...
```

The commands **help** or ? supply a more detailed description of the library:

```
>> ?numlib

   numlib -- Number theory

   Description:

   This documentation describes the routines of the
   library numlib. This library contains the following
   routines:

   decimal
        infinite representation of rational numbers
   divisors
        divisors of an integer
   fibonacci
        Fibonacci numbers
   ...
```

If you have a graphical user interface, then this command opens a separate help window, and you can navigate to the help page of one of the listed functions by clicking on its name with the mouse. In MuPAD versions up to 1.4, you can obtain a list of all available libraries by choosing the menu entry "Helpindex" (which resides in the "Targets" menu on a Windows platform) and then clicking on "Library Packages". In MuPAD versions beyond 1.4, the command **info()** also provides such a list.

The function **numlib::decimal** of the **numlib** library computes the decimal expansion of a rational number[1]:

```
>> numlib::decimal(123/7);

   17, [5, 7, 1, 4, 2, 8]
```

As for other system functions, you may request information on library functions by means of **help** or ?:

---

[1] The following result is to be interpreted as follows: $123/7 = 17.\overline{571428} = 17.571428\,571428\ldots$

```
>> ?numlib::decimal
```

You can have a look at the implemention of a library function by using expose:

```
>> expose(numlib::fibonacci);

  proc(n)
    name numlib::fibonacci;
    local x, y, z, a, b, c, Z, C;
  begin
    if testargs() then
      ...
  end_proc
```

3.2 Exporting Libraries

You have seen in the previous section that the call syntax for a library function is library::function, where library and function are the names of the library and the function, respectively. For example, the library numeric for numerical computations contains the function numeric::newton[2]. It implements the well-known Newton method for numerical root finding. In the following example, we approximate a root of the sine function in the interval $[2, 4]$:

```
>> numeric::newton(sin(x), x = 2..4);

  3.141592653
```

The function export makes functions of a library "globally known", so that you can use them without specifying the library name:

```
>> export(numeric, newton): newton(sin(x), x = 2..4);

  3.141592653
```

If you already have assigned a value to newton, then export returns an error message:

---

[2] The function numeric::newton is replaced by numeric::solve in MuPAD versions beyond 1.4.

```
>> newton := 1: export(numeric, newton);
```

> Error: global name must be identifier [export]

You can export several functions at once:

```
>> export(numeric, fsolve, quadrature):
```

Now you can use fsolve[3] (to find *all* real roots of a polynomial) and quadrature (for numerical integration) directly. We refer to the corresponding help pages for the meaning of the input parameters and the returned output.

```
>> fsolve(x^3 - 6*x^2 + 11*x - 6, x, -10, 10, 0.001);
```

> 255/256..8225/8192, 16265/8192..4125/2048,
>
>   765/256..24665/8192

```
>> quadrature(exp(x) + 1, x = 0..1);
```

>   2.718281828

If you call export with only one argument, namely the name of the library, then all functions in that library are exported. If there are name conflicts with already existing identifiers, then export issues warnings:

```
>> eigenvalues := 1: export(numeric);
```

> Warning: 'newton' already is exported
> Warning: 'eigenvalues' already has a value, not expor\
> ted
> Warning: 'quadrature' already is exported
> Warning: 'fsolve' already is exported

After deleting the identifier, the library function with the same name can be exported successfully:

```
>> unassign(eigenvalues): export(numeric, eigenvalues):
```

---

[3] The function numeric::fsolve is replaced by numeric::realroots in MuPAD versions beyond 1.4.

3.3 The Standard Library

MuPAD's most important library is the standard library `stdlib`. It contains the most frequently used functions such as `diff`, `simplify`, etc. All functions from `stdlib` are exported automatically at startup. In this respect, there is no notable difference between MuPAD's kernel functions, which are written in C, and the functions from the standard library, which are implemented in MuPAD's programming language.

If you execute MuPAD on a graphical user interface, then you can obtain more information about the available functions of the standard library by choosing "Helpindex" (which resides in the "Targets" menu on Windows platforms) in the help window and then clicking on "Standard Library". The MuPAD Quick Reference [Oev 98] lists all functions of the kernel and the standard library in MuPAD version 1.4.

Many of these functions are implemented as function environments (Section 18.11). You can view the source code via `expose(name)`:

```
>> expose(exp);

   proc(x)
     name exp;
     local y;
   begin
     if x::exp <> FAIL then
       return(x::exp(args()))
     end_if;
     ...
   end_proc
```

4. **MuPAD** Objects

In Chapter 2, we have introduced MuPAD objects such as numbers, symbolic expressions, maps, or matrices. Now we discuss these objects more systematically.

The objects sent to the kernel for evaluation can be of various shapes: simple arithmetic expressions with numbers such as 1+(1+I)/3, arithmetic expressions with symbolic objects such as x+(y+I)/3, lists, sets, equations, inequalities, maps, arrays, abstract mathematical objects, and more. Every MuPAD object belongs to some data type, called the *domain type*. It corresponds to a certain internal representation of the object. In what follows, we discuss the following fundamental domain types:

domain type	meaning
DOM_INT	integers, e.g., -3, 10^5
DOM_RAT	rational numbers, e.g., 7/11
DOM_FLOAT	floating-point numbers, e.g., 0.123
DOM_COMPLEX	complex numbers, e.g., 0.1 + 2/3*I
DOM_IDENT	symbolic identifiers, e.g., x, y, f
DOM_EXPR	symbolic expressions, e.g., x + y
Puiseux	symbolic series expansions, e.g., 1 + x + x^2 + O(x^3)
DOM_LIST	lists, e.g., [1, 2, 3]
DOM_SET	sets, e.g., {1,2,3}
DOM_ARRAY	arrays
DOM_TABLE	tables
DOM_BOOL	Boolean values: TRUE, FALSE, UNKNOWN
DOM_STRING	strings, e.g., "I am a string"
Dom::Matrix(..)	matrices and vectors
DOM_POLY	polynomials, e.g., poly(x^2 + x + 1, [x])
DOM_PROC	functions and procedures

Moreover, you can define your own data types, but we do not discuss this here[1]. The system function **domtype** returns the domain type of a MuPAD object.

In the following section, we first present the important operand function **op**, which enables you to decompose a MuPAD object into its building blocks. The following sections discuss the above data types and some of the main system functions for handling them.

4.1 Operands: the Functions op and nops

It is often necessary to decompose a MuPAD object into its components in order to process them individually. The building blocks of an object are called *operands*. The system functions for accessing them are **op** and **nops** (short for: number of operands):

nops(object)   : the number of operands,

op(object,i)   : the ith operand, $0 \leq i \leq$ nops(object)

op(object,i..j) : the sequence of operands i through j,
                  where $0 \leq i \leq j \leq$ nops(object),

op(object)   : the sequence op(.,1),op(.,2),.. of all
                  operands.

The meaning of an operand depends on the data type of the object. We discuss this for each data type in detail in the following sections. For example, the operands of a rational number are the numerator and the denominator, the operands of a list or set are the elements, and the operands of a function call are the arguments. However, there are also objects where the decomposition into operands is less intuitive, such as series expansions as generated by the system functions **taylor** or **series** (Section 4.13). Here is an example with a list (Section 4.6):

>> list := [a, b, c, d, sin(x)]: nops(list);

5

---

[1] The interested reader may consult the MuPAD manual [MuP 96]. You find a simple example in the document [DPS 97], which is available in the online help system via ?advdemo. More detailed information is contained in [Dre 97].

```
>> op(list, 2);
```

   b

```
>> op(list, 3..5);
```

   c, d, sin(x)

```
>> op(list);
```

   a, b, c, d, sin(x)

By repeatedly calling the op function, you can decompose arbitrary
MuPAD expressions into "atomic" ones. In this model, a MuPAD
atom is an expression that cannot be further decomposed by op, such
that op(atom)=atom holds[2]. This is essentially the case for integers,
floating-point numbers, identifiers that have not been assigned a value,
and strings:

```
>> op(-2), op(0.1234), op(a), op("I am a text");
```

   -2, 0.1234, a, "I am a text"

In the following example, a nested list is decomposed completely into
its atoms a11,a12,a21,x,2:

```
>> list := [[a11, a12], [a21, x^2]];
```

The operands and suboperands are:

op(list, 1)	:	[a11, a12]
op(list, 2)	:	[a21, x^2]
op(op(list, 1), 1)	:	a11
op(op(list, 1), 2)	:	a12
op(op(list, 2), 1)	:	a21
op(op(list, 2), 2)	:	x^2
op(op(op(list, 2), 2), 1)	:	x
op(op(op(list, 2), 2), 2)	:	2

---

[2] This model is a good approximation to MuPAD's internal mode of operation, but
there are some exceptions. For example, you can decompose rational numbers via
op, but the kernel regards them as atoms. On the other hand, although strings
are indecomposable with respect to op, it is still possible to access the characters
of a string individually (Section 4.11).

Instead of the annoying nested calls of op, you may also use the following short form to access subexpressions:

```
op(list, [1])         :    [a11, a12]
op(list, [2])         :    [a21, x^2]
op(list, [1, 1])      :    a11
op(list, [1, 2])      :    a12
op(list, [2, 1])      :    a21
op(list, [2, 2])      :    x^2
op(list, [2, 2, 1])   :    x
op(list, [2, 2, 2])   :    2
```

Exercise 4.1: Determine the operands of the power a^b, the equation a=b, and the symbolic function call f(a,b).

Exercise 4.2: The following call of solve (Chapter 8) returns a set:

```
>> set := solve({x + sin(3)*y = exp(a),
                 y - sin(3)*y = exp(-a)}, {x,y});
```

```
{ {                    sin(3) exp(-a)          exp(-a)   } }
{ { x = exp(a) -   ---------------,  y =  ----------- } }
{ {                    1 - sin(3)          1 - sin(3) } }
```

Extract the value of the solution for y and assign it to the identifier y.

4.2 Numbers

We have demonstrated in Section 2.2 how to work with numbers. There are various data types for numbers:

```
>> domtype(-10), domtype(2/3), domtype(0.1234),
   domtype(0.1 + 2*I);

   DOM_INT, DOM_RAT, DOM_FLOAT, DOM_COMPLEX
```

A rational number is a compound object: the building blocks are the numerator and the denominator. Similarly, a complex number consists of the real and the imaginary part. You can use the operand function op from the previous section to access these components:

```
>> op(111/223, 1), op(111/223, 2);

   111, 223

>> op(100 + 200*I, 1), op(100 + 200*I, 2);

   100, 200
```

Alternatively, you can use the system functions **numer**, **denom**, **Re**, and **Im**:

```
>> numer(111/223), denom(111/223),
   Re(100 + 200*I), Im(100 + 200*I);

   111, 223, 100, 200
```

Besides the common arithmetic operations +, −, *, and /, there are the arithmetic operators **div** and **mod**, for division of an integer x by a nonzero integer p with remainder. If $x = k\,p + r$ holds with integers k and $0 \le r < |p|$, then x **div** p returns the "integral quotient" k and x **mod** p returns the "remainder" r:

```
>> 25 div 4, 25 mod 4;

   6, 1
```

Table 4.1 gives a compilation of the main MuPAD functions and operators for handling numbers. We refer to the help system (i.e., ?abs, ?ceil, etc.) for a detailed description of these functions. We stress that while expressions such as $\sqrt{2}$ mathematically represent numbers, MuPAD treats them as symbolic expressions (Section 4.4):

```
>> domtype(sqrt(2));

   DOM_EXPR
```

Exercise 4.3: What is the difference between $1/3 + 1/3 + 1/3$ and $1.0/3 + 1/3 + 1/3$ in MuPAD?

`+, -, *, /, ^`	:	basic arithmetic
`abs`	:	absolute value
`ceil`	:	rounding up
`div`	:	quotient "modulo"
`fact`	:	factorial
`float`	:	approximation by floating-point numbers
`floor`	:	rounding down
`frac`	:	fractional part
`ifactor, Factor`	:	prime factorization
`isprime`	:	primality test
`mod`	:	remainder "modulo"
`round`	:	rounding
`sign`	:	sign
`sqrt`	:	square root
`trunc`	:	integral part

Table 4.1. MuPAD functions and operators for numbers.

Exercise 4.4: Compute the decimal expansions of $\pi^{(\pi^\pi)}$ and $e^{\frac{1}{3}\pi\sqrt{163}}$ with a precision of 10 and 100 digits, respectively. What is the 234th digit after the decimal point of π?

Exercise 4.5: After you execute `x:=10^50/3.0`, only the first `DIGITS` decimal digits of `x` are guaranteed to be correct.

a) Truncating the fractional part via `trunc` is therefore questionable. What does MuPAD do?
b) What is returned for `x` after increasing `DIGITS`?

4.3 Identifiers

Identifiers are names, such as `x` or `f`, that may represent variables and unknowns. They may consist of arbitrary combinations of letters, digits, and the underscore "`_`", with the only exception that the first symbol must not be a digit. MuPAD distinguishes uppercase and lowercase letters. Examples of admissible identifiers are `x`, `_x23`, and `the_MuPAD_system`, while MuPAD would not accept `12x`, `p-2`, and `x>y` as identifiers.

Identifiers that have not been assigned a value evaluate to their name. In MuPAD, they represent symbolic objects such as unknowns in equations. Their domain type is `DOM_IDENT`:

```
>> domtype(x);
```

    DOM_IDENT

You can assign an arbitrary object to an identifier by means of the *assignment operator* :=. Afterwards, this object is the *value* of the identifier. For example, after the command

```
>> x := 1 + I:
```

the identifier x has the value 1+I, which is a complex number of domain type DOM_COMPLEX. You should be careful to distinguish between an identifier, its value, and its evaluation. We refer to the important Chapter 5, where MuPAD's evaluation strategy is described.

If an identifier already has been assigned a value, then another assignment overwrites the previous value. The statement y:=x does not assign the identifier x to the identifier y, but the current value (the evaluation) of x:

```
>> x := 1: y := x: x, y;
```

    1, 1

If the value of x is changed later on, then this does not affect y:

```
>> x := 2: x, y;
```

    2, 1

However, if x is a symbolic identifier, which evaluates to itself, then the new identifier y refers to this symbol:

```
>> unassign(x): y := x: x, y; x := 2: x, y;
```

    x, x

    2, 2

Here we have deleted the value of the identifier x by means of the function **unassign**[3], and x has become a symbolic identifier without a value again.

The assignment operator := is a short form of the system function **_assign**, which may also be called directly:

```
>> _assign(x, value): x;

    value
```

This function returns its second argument, namely the right hand side of an assignment. This explains the screen output after an assignment:

```
>> y := 2*x;

    2 value
```

You can work with the returned value immediately. For example, the following construction is allowed (the assignment must be put in parenthesis):

```
>> y := cos( (x := 0) ): x, y;

    0, 1
```

Here the value 0 is assigned to the identifier x. The return value of the assignment, i.e., 0, is fed directly as argument to the cosine function, and the result $\cos(0) = 1$ is assigned to y. Thus we have simultaneously assigned values to both x and y.

Another assignment function in MuPAD is **assign**. Its input are sets or lists of equations, which are transformed into assignments:

```
>> unassign(x, y): assign({x = 0, y = 1}): x, y;

    0, 1
```

This function is particularly useful in connection with **solve** (Section 8), which returns solutions as sets of equations of the form **identifier=value**, without assigning these values.

---

[3] The keyword **delete** replaces the function **unassign** in MuPAD versions beyond 1.4.

There exist many identifiers in MuPAD with predefined values. They represent mathematical functions (such as **sin**, **exp**, or **sqrt**), mathematical constants (such as PI), or MuPAD algorithms (such as **diff**, **int**, or **limit**). If you try to change the value of such a predefined identifier, then MuPAD issues a warning or an error message:

```
>> sin := 1;
```

```
  Error: Illegal assignment. Identifier is
         write protected [sin]
```

You can protect you own identifiers against overwriting via the command **protect(identifier)**. The write protection of both your own and the system identifiers can be removed by **unprotect(identifier)**. However, we strongly recommend not to overwrite predefined identifiers since they are used by many system functions which would return unpredictable results after a redefinition. The command[4] **anames(3)** lists all predefined identifiers.

You can use the concatenation operator "**.**" to generate names of identifiers dynamically. If **x** and **i** are identifiers, then **x.i** generates a new identifier by concatenating the *evaluations* (see Chapter 5) of **x** and **i**:

```
>> x := z: i := 2:  x.i;
```

```
  z2
```

```
>> x.i := value: z2;
```

```
  value
```

In the following example, we use a **for** loop (Chapter 16) to assign values to the identifiers **x1,..,x1000**:

```
>> unassign(x):
>> for i from 1 to 1000 do x.i := i^2 end_for:
```

Due to possible side effects or conflicts with already existing identifiers, we strongly recommend to use this concept only interactively and not within MuPAD procedures.

---

[4] The corresponding command in MuPAD versions beyond 1.4 is **anames(All)**.

The function **genident** generates a new identifier that has not been used before in a MuPAD session:

```
>> X3 := (X2 := (X1 := 0)): genident();

    X4
```

You may use strings enclosed in quotation marks " (Section 4.11) to generate identifiers dynamically:

```
>> a := email: b := "4you": a.b;

    email4you
```

However, there are some restrictions, since the resulting name must be a valid MuPAD identifier. In particular, the string must not contain blanks or arithmetical symbols:

```
>> a := email: b := "4you + x": a.b;

    Error: Illegal argument [_concat]
```

Strings are not identifiers and cannot be assigned a value:

```
>> "string" := 1;

    Syntax Error: Unexpected symbol in assignment

    "string" := 1;
             ^
```

Exercise 4.6: Which of the following names x, x2, 2x, x_t, diff, exp, caution!-!, x-y, Jack&Jill, a_valid_identifier are valid identifiers in MuPAD? Which of them can be assigned values?

Exercise 4.7: Read the help page for **solve**. Solve the system of equations

$$x_1 + x_2 = 1 \ , \ x_2 + x_3 = 1 \ , \ \ldots \ , \ x_{19} + x_{20} = 1 \ , \ x_{20} = \pi$$

in the unknowns $x_1, x_2, \ldots, x_{20}$. Read the help page for **assign** and assign the values of the solution to the unknowns.

4.4 Symbolic Expressions

We say that an object containing symbolic terms, such as the equation

$$0.3 + \sin(3) + \frac{f(x,y)}{5} = 0 \ ,$$

is an *expression*. Expressions of domain type DOM_EXPR are probably the most general data type in MuPAD. Expressions are built of atomic components, as all MuPAD objects, and are composed by means of *operators*. This comprises binary operators, such as the basic arithmetic operations $+, -, *, /, \char`^$, and function calls such as sin(..), f(..), etc.

4.4.1 Operators

MuPAD throughout uses functions to combine or manipulate objects[5]. It would be little intuitive, however, to use function calls everywhere, say, _plus(a,b) for the addition $a + b$. For that reason, a variety of important operations is implemented in such a way that you can use the familiar mathematical notation ("operator notation") for input. Also the output is given in such a form. In the following, we list the operators for building more complex MuPAD expressions from atoms.

The operators +, -, *, / for the basic arithmetic operations and $\char`^$ for exponentiation are valid for symbolic expressions as well:

```
>> a + b + c, a - b, a*b*c, a/b, a^b;
```

$$a + b + c, \ a - b, \ a\ b\ c, \ \frac{a}{b}, \ a^b$$

You may input these operators in the familiar mathematical way, but internally they are function calls:

```
>> _plus(a,b,c), _subtract(a,b), _mult(a,b,c),
   _divide(a,b), _power(a,b);
```

---

[5] Remarkably, the MuPAD kernel treats not only genuine function calls, such as sin(0.2), assignments, or arithmetical operations in a functional way, but also loops (Chapter 16) and case distinctions (Chapter 17).

$$a + b + c, \ a - b, \ a \, b \, c, \ -\frac{a}{b}, \ a$$

The same holds for the factorial of a nonnegative integer. You may
input it in the mathematical notation n!. Internally it is converted to
a call of the function fact:

```
>> n!, fact(10);
```

```
   fact(n), 3628800
```

The arithmetic operators div and mod[6] have been presented in Chapter 4.2. They may also be used in a symbolic context, but then return
only symbolic results:

```
>> x div 4, 25 mod p;
```

```
   x div 4, 25 mod p
```

Several MuPAD objects separated by commas form a sequence:

```
>> sequence := a, b, c + d;
```

```
   a, b, c + d
```

The operator $ is an important tool to generate such sequences:

```
>> i^2 $ i = 2..7 ;   x^i $ i = 1..5;
```

```
   4, 9, 16, 25, 36, 49
```

```
        2   3   4   5
   x, x , x , x , x
```

Equations and inequalities are valid MuPAD objects. They are generated by the equality sign = and by <>, respectively:

---

[6] The object x mod p is internally converted to the function call _mod(x,p). The
function _mod can be redefined, e.g., _mod := modp or _mod := mods. The behavior of modp and mods is documented on the corresponding help pages. A redefinition of _mod also redefines the operator mod.

```
>> equation := x + y = 2; inequality := x <> y;
```

```
   x + y = 2
```

```
   x <> y
```

The operators <, <=, >, and >= compare the magnitudes of their arguments. The corresponding expressions represent conditions:

```
>> condition := i <= 2;
```

```
   i <= 2
```

In a concrete context, they usually can be evaluated to one of the
truth ("Boolean") values **TRUE** or **FALSE**. Typically, they are used in
if statements or as termination conditions in loops. You may combine
Boolean expressions via the logical operators **and** and **or**, or negate
them via **not**:

```
>> condition3 := condition1 and (not condition2);
```

```
   condition1 and not condition2
```

You can define maps (functions) in several ways in MuPAD. The
simplest method is to use the *arrow operator* -> (the minus symbol
followed by the "greater than" symbol):

```
>> f := x -> (x^2);
```

```
        2
   x -> x
```

The parentheses around the right hand is obsolete in MuPAD versions
beyond 1.4. After defining a function in this way, you may call it like
a system function:

```
>> f(4), f(x + 1), f(y);
```

```
           2    2
   16, (x + 1) , y
```

The composition of functions if defined by means of the *composition operator* @:

```
>> c := a@b: c(x);
```

a(b(x))

The *iteration operator* @@ is for iterated composition of a function with itself:

```
>> f := g@@4: f(x);
```

g(g(g(g(x))))

Some system functions, such as definite integration via int or the $ operator, require a *range*. You generate a range by means of the operator ..:

```
>> range := 0..1;  int(x, x = range);
```

0..1

1/2

MuPAD treats any expression of the form identifier(argument) as a function call:

```
>> unassign(f):
   expression := sin(x) + f(x, y) + int(g(x), x = 0..1);
```

sin(x) + f(x, y) + int(g(x), x = 0..1)

Table 4.2 lists all operators presented above, together with their equivalent functional form. You may use either form to input expressions:[7]

```
>> 2/14 = _divide(2, 14),
   [i $ i = 3..5] = [_seqgen(i, i = 3..5)];
```

1/7 = 1/7, [3, 4, 5] = [3, 4, 5]

---

[7] In MuPAD versions beyond 1.4, the syntax of the function _seqgen changes; see the footnote on page 68.

operator	system function	meaning	example
+	_plus	addition	SUM := a+b
−	_subtract	subtraction	Difference := a-b
*	_mult	multiplication	Product := a*b
/	_divide	division	Quotient := a/b
^	_power	exponentiation	Power := a^b
div	_div	quotient modulo p	Quotient := a div p
mod	_mod	remainder modulo p	Remainder := a mod p
!	fact	factorial	n!
$	_seqgen	sequence generation	Sequence := i^2 $ i=3..5
,	_exprseq	sequence concatenation	Seq := Seq1,Seq2
union	_union	set union	S := Set1 union Set2
intersect	_intersect	set intersection	S := Set1 intersect Set2
minus	_minus	set difference	S := Set1 minus Set2
=	_equal	equation	Equation := x+y=2
<>	_unequal	inequality	Condition := x<>y
<	_less	comparison	Condition := a<b
>		comparison	Condition := a>b
<=	_leequal	comparison	Condition := a<=b
>=		comparison	Condition := a>=b
not	_not	negation	Condition2 := not Condition1
and	_and	logical 'and'	Condition := a<b and b<c
or	_or	logical 'or'	Condition := a<b or b<c
->		mapping	Square := x -> (x^2)
'	D	differential operator	f'(x)
@	_fconcat	composition	h := f@g
@@	repcom	iteration	g := f@@3
..	_range	range	Range := a..b
.	_concat	concatenation	NewName := Name1.Name2
identifier()		function call	sin(1), f(x), reset()

Table 4.2. The main operators for generating MuPAD expressions

---

```
>> a < b = _less(a, b), (f@g)(x) = _fconcat(f, g)(x);

   (a < b) = (a < b), f(g(x)) = f(g(x))
```

We remark that some of the system functions, such as _plus, _mult, _union, or _concat, accept arbitrarily many arguments, while the corresponding operators are only binary:

```
>> _plus(a, b, u, v), _concat(a, b, u, v);
```

```
   a + b + u + v, abuv
```

It is often useful to know and to use the functional form of the operators. For example, it is very efficient to form longer sums by applying
_plus to many arguments. You may generate the argument sequence
quickly by means of the sequence generator $:

```
>> _plus(1/i! $ i = 0..100): float(%);
```

```
   2.718281828
```

The function map is a useful tool. It applies a function to all
operands of a MuPAD object. For example, the call

```
>> map([x1, x2, x3], function, y, z);
```

returns the list (Section 4.6):

```
   [function(x1, y, z), function(x2, y, z),

     function(x3, y, z)]
```

If you want to apply operators via map, use their functional equivalent[8]:

```
>> map([x1, x2, x3], _power, 5), map([f, g], repcom, 5);
```

```
        5    5    5
   [x1 , x2 , x3 ], [f@f@f@f@f@f, g@g@g@g@g@g]
```

Some operations are invalid if they do not make sense mathematically:

```
>> 3 and x;
```

```
   Error: Illegal operand [_and]
```

---

[8] The functional equivalent repcom (short for: repeated composition) of the iteration operator @@ is denoted by _fnest in MuPAD versions beyond 1.4.

The system function _and recognizes that the argument 3 cannot represent a Boolean value, and issues an error message. However, MuPAD accepts a symbolic expression such as a and b, with symbolic identifiers a,b. As soon as a and b get Boolean values, the expression can be evaluated to a Boolean value as well:

```
>> c := a and b: a := TRUE: b := TRUE: c;

   TRUE
```

The operators have different *priorities*, for example:

a.fact(3)	means a.(fact(3)) and returns a6,
a.6^2	means (a.6)^2 and returns a6^2,
a*b^c	means a*(b^c),
a + b*c	means a + (b*c),
a + b mod c	means (a + b) mod c,
a = b mod c	means a = (b mod c),
a, b $ 3	means a, (b $ 3) and returns a, b, b, b.

If we denote the relation "is of lower priority than" by $\prec$, then we have:

$$, \prec \$ \prec = \prec \mbox{mod} \prec + \prec * \prec \hat{} \prec . \prec \mbox{function call.}$$

You find a complete list of the operators and their priorities in Section "Operators" of the MuPAD Quick Reference [Oev 98]. Parentheses can always be used to the priority:

```
>> 1 + 1 mod 2 ,   1 + (1 mod 2);

   0, 2

>> i := 2: x.i^2, x.(i^2);

     2
   x2 , x4

>> u, v $ 3 ;   (u, v) $ 3;

   u, v, v, v


   u, v, u, v, u, v
```

4.4.2 Expression Trees

A useful model for representing a MuPAD expression is the *expression tree*. It reflects the internal representation. The operators or their corresponding functions, respectively, are the vertices, and the arguments are subtrees. The operator of lowest priority is at the root. Here are some examples:

```
a + b*c + d*e*sin(f)^g
```

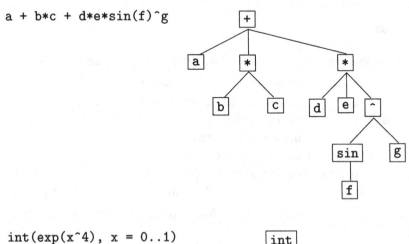

```
int(exp(x^4), x = 0..1)
```

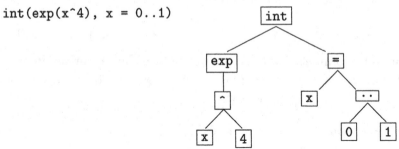

The difference **a-b** is internally represented as **a+b*(-1)**:

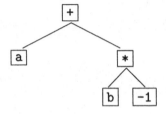

Similarly, a quotient **a/b** has the internal representation **a*b^(-1)**:

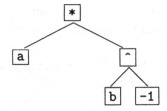

The leaves of an expression tree are MuPAD atoms.

Exercise 4.8: Sketch the expression tree of a^b-sin(a/b).

Exercise 4.9: Determine the operands of 2/3, x/3, 1+2*I, and x+2*I. Explain the differences that you observe.

4.4.3 Operands

You can decompose expressions systematically by means of the operand functions op and nops, which we have already presented in Section 4.1. The operands of an expression correspond to the subtrees below the root in the expression tree.

```
>> expression := a + b + c + sin(x): nops(expression);

   4

>> op(expression);

   a, b, c, sin(x)
```

Additionally, expressions of domain type **DOM_EXPR** have a "0th operand", which is accessible via op(..,0). It corresponds to the root of the expression tree and tells you which system function is used to build the expression:

```
>> op(a + b*c, 0), op(a*b^c, 0), op(a^(b*c), 0);

   _plus, _mult, _power

>> sequence := a, b, c:  op(sequence, 0);

   _exprseq
```

If the expression is an unevaluated function call, then the command op(expression,0) returns the identifier of that function:

```
>> op(sin(1), 0), op(f(x), 0), op(diff(y(x), x), 0),
   op(int(exp(x^4), x), 0);

   sin, f, diff, int
```

You may regard the 0th operand of an expression as a "mathematical type". For example, an algorithm for differentiation of arbitrary expressions must find out whether the expression is a sum, a product, or a function call. To this end, it may look at the 0th operand to decide whether linearity, the product rule, or the chain rule of differentiation applies.

As an example, we decompose the expression

```
>> expression := a + b + sin(x) + c^2:
```

with the expression tree

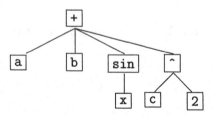

systematically by means of the op function:

```
>> op(expression, 0..nops(expression));

                    2
   _plus, a, b, sin(x), c
```

We can put these expressions together again in the following form:

```
>> root := op(expression, 0): operands := op(expression):
>> root(operands);

                  2
   a + b + sin(x) + c
```

In the following example, we decompose an expression completely into its atoms x, a, b (compare with Section 4.1):

```
>> expression := sin(x + cos(a*b)):
```

The operands and subexpressions are:

op(expression, 0)	: sin
op(expression, 1)	: x+cos(a*b)
op(expression, [1, 0])	: _plus
op(expression, [1, 1])	: x
op(expression, [1, 2])	: cos(a*b)
op(expression, [1, 2, 0])	: cos
op(expression, [1, 2, 1])	: a*b
op(expression, [1, 2, 1, 0])	: _mult
op(expression, [1, 2, 1, 1])	: a
op(expression, [1, 2, 1, 2])	: b

Exercise 4.10: Sketch the expression tree of the following Boolean expression:

```
>> condition := (not a) and (b or c):
```

How can you use op to pick the symbolic identifiers a, b, and c out of the object condition?

4.5 Sequences

Sequences form an important data structure in MuPAD. Lists and sets are built from sequences. As discussed in Section 4.4, a sequence is a series of MuPAD objects separated by commas.

```
>> sequence1 := a, b, c; sequence2 := c, d, e;
```

```
    a, b, c
```

```
    c, d, e
```

You may also use the comma to concatenate sequences:

```
>> sequence3 := sequence1, sequence2;
```

    a, b, c, c, d, e

Sequences are MuPAD expressions of domain type DOM_EXPR. If m and n are integers, then the call object(i) $ i=m..n generates the sequence

$$object(m), \ object(m+1), \ \ldots, \ object(n) :$$

```
>> i^2 $ i = 2..7 , x^i $ i = 1..5;
```

                            2   3   4   5
    4, 9, 16, 25, 36, 49, x, x , x , x , x

The operator $ is called the *sequence generator*. The equivalent functional form is _seqgen(object(i),i=m..n)[9]:

```
>> _seqgen(i^2, i = 2..7) , _seqgen(x^i, i = 1..5);
```

                            2   3   4   5
    4, 9, 16, 25, 36, 49, x, x , x , x , x

Usually, you will prefer the operator notation; the functional form is useful in connection with map, zip, or similar functions. In MuPAD version 1.4, the loop variable must be an identifier without a value, otherwise an error occurs:

```
>> i := 1: i^2 $ i = 2..7;
```

    Error: Illegal argument [_seqgen]

In MuPAD versions beyond 1.4 loop variables may already have a value[10]. You may use $ in the following way to generate a sequence of successive integers:

---

[9] In MuPAD versions beyond 1.4, the syntax is slightly different: you have to use _seqgen(object(i),i,m..n) instead.

[10] In version 1.4 it is useful to enclose the loop variable on the right hand side of the $ operator via hold (Section 5.2):

```
>> i^2 $ hold(i) = 2..7;
```

In versions beyond 1.4, hold is no longer allowed in a $ call.

```
>> $ 23..30;
```

```
   23, 24, 25, 26, 27, 28, 29, 30
```

The command `object $ n` returns sequence of n identical objects:

```
>> x^2 $ 10;
```

$$x^2, x^2, x^2, x^2, x^2, x^2, x^2, x^2, x^2, x^2$$

You can also use the sequence generator in connection with the keyword `in`. The loop variable then runs through all operands of the stated object:

```
>> f(x) $ x in [a, b, c, d];
```

```
   f(a), f(b), f(c), f(d)
```

```
>> f(x) $ x in a + b + c + d + sin(sqrt(2));
```

$$f(a), f(b), f(c), f(d), f(\sin(2^{1/2}))$$

It is easy to let MuPAD execute a sequence of commands by means of `$`. In the following example two assignments (separated by semicolons) are performed in each step. Afterwards, the identifiers have the corresponding values:

```
>> unassign(i): (x.i := sin(i); y.i := x.i) $ i=1..4:
>> x1, x2, y3, y4;
```

```
   sin(1), sin(2), sin(3), sin(4)
```

As a simple application of sequences, we now consider the MuPAD differentiator `diff`. The call `diff(f(x),x)` returns the derivative of f with respect to x. You may compute higher derivatives by `diff(f(x),x,x)`, `diff(f(x),x,x,x)`, etc. Thus the 10th derivative of $f(x) = \sin(x^2)$ can be computed conveniently by means of the sequence generator:

```
>> diff(sin(x^2), x $ 10);
```

$$30240 \cos(x^2) - 403200 x^4 \cos(x^2) - 302400 x^2 \sin(x^2)$$

$$+ 23040 x^8 \cos(x^2) + 161280 x^6 \sin(x^2) -$$

$$1024 x^{10} \sin(x^2)$$

MuPAD's "void" object (Section 4.17) may be regarded as an empty sequence. You may generate it by calling `null()` or `_exprseq()`. The system automatically eliminates this object from sequences:

```
>> Seq := null(): Seq := Seq, a, b, null(), c;
```

```
   a, b, c
```

Some system functions, such as the `print` command for screen output (Section 13.1.1), return the `null()` object:

```
>> sequence := a, b, print(Hello) , c;
```

```
            Hello
```

```
   a, b, c
```

You can access the *i*th entry of a sequence by `sequence[i]`. Redefinitions of the form `sequence[i]:=newvalue` are also possible:

```
>> F := a, b, c: F[2];
```

```
   b
```

```
>> F[2] := newvalue: F;
```

```
   a, newvalue, c
```

Alternatively, you may use the operand function **op** (Section 4.1) to access subsequences[11]:

```
>> F := a, b, c, d, e: op(F, 2); op(F, 2..4);
```

b

b, c, d

You may use **unassign**[12] to delete entries from a sequence, thus shortening the sequence:

```
>> F; unassign(F[2]): F; unassign(F[3]): F;
```

a, b, c, d, e

a, c, d, e

a, c, e

The main usage of sequences in MuPAD is the generation of lists and sets and supplying arguments to function calls. For example, the functions **max** and **min** can compute the maximum and minimum, respectively, of arbitrarily many arguments:

```
>> Seq := 1, 2, -1, 3, 0: max(Seq), min(Seq);
```

3, -1

Exercise 4.11: Assign the values $x_1 = 1$, $x_2 = 2$, ... , $x_{100} = 100$ to the identifiers $x_1, x_2, \ldots, x_{100}$.

---

[11] Note that in this example the identifier F of the sequence is provided as argument to op. The op function regards a direct call of the form op(a,b,c,d,e,2) as an (invalid) call with six arguments and issues an error message. You may use parentheses to avoid this error: op((a,b,c,d,e,2)).

[12] The keyword **delete** replaces the function **unassign** in MuPAD versions beyond 1.4.

Exercise 4.12: Generate the sequence

$$x_1, \underbrace{x_2, x_2}_{2}, \underbrace{x_3, x_3, x_3}_{3}, \ldots, \underbrace{x_{10}, x_{10}, \ldots, x_{10}}_{10} \,.$$

Exercise 4.13: Use a simple command to generate the double sum

$$\sum_{i=1}^{10} \sum_{j=1}^{i} \frac{1}{i+j} \,.$$

Hint: the function _plus accepts arbitrarily many arguments. Generate a suitable argument sequence.

4.6 Lists

As list is an ordered sequence of arbitrary MuPAD objects enclosed in square brackets:

```
>> list := [a, 5, sin(x)^2 + 4, [a, b, c], hello,
            3/4, 3.9087];
```

```
                  2
    [a, 5, sin(x)  + 4, [a, b, c], hello, 3/4, 3.9087]
```

A list may contain lists as elements. It may be empty:

```
>> list := [ ];
```

```
    []
```

The possibility to generate sequences via $ is helpful for constructing lists:

```
>> sequence := i $ i = 1..10 : list := [sequence];
```

```
    [1, 2, 3, 4, 5, 6, 7, 8, 9, 10]
```

```
>> list := [x^i $ i = 0..12];
```

```
       2   3   4   5   6   7   8   9   10   11   12
[1, x, x , x , x , x , x , x , x , x  , x  , x  ]
```

A list may occur on the left hand side of an assignment. This may be used to assign values to several identifiers simultaneously:

```
>> ([A, B, C]) := [a, b, c]: A + B^C;
```

```
         c
    a + b
```

The function **nops** returns the number of elements of a list. You can access the elements of a list by means of the **op** function: op(list) returns the sequence of elements, i.e., the sequence that has been used to construct the list by enclosing it in square brackets [ ]. The call op(list,i) returns the ith list element, and op(list,i..j) extracts the sequence of the ith up to the jth list element:

```
>> list := [a, b, sin(x), c]: op(list);
```

```
    a, b, sin(x), c
```

```
>> op(list, 2..3);
```

```
    b, sin(x)
```

The index operator provides an alternative way of accessing list elements[13]:

```
>> list[1], list[2];
```

```
    a, b
```

You may change a list element by an indexed assignment:

```
>> list := [a, b, c]: list[1] := newvalue: list;
```

```
    [newvalue, b, c]
```

---

[13] This method of accessing list elements is faster than using the function op.

Alternatively, the command subsop(list,i=newvalue) (Chapter 6) redefines the ith operand:

```
>> list := [a, b, c]: list := subsop(list, 1 = newvalue);

   [newvalue, b, c]
```

Caution: If L is an identifier without a value, then the indexed assignment

```
>> L[index] := value:
```

generates a table (Section 4.8) and not a list:

```
>> unassign(L): L[1] := a: L;

   table(
     1 = a
   )
```

You can remove elements from a list by using **unassign**[14]. This shortens the list:

```
>> list := [a, b, c]: unassign(list[1]): list;

   [b, c]
```

The function contains checks whether a MuPAD object belongs to a list. It returns the index of (the first occurrence of) the element in the list. If the list does not contain the element, then contains returns the integer 0:

```
>> contains([x + 1, a, x + 1], x + 1);

   1

>> contains([sin(a), b, c], a);

   0
```

---

[14] The keyword **delete** replaces the function **unassign** in MuPAD versions beyond 1.4.

The function **append** adds elements to the tail of a list:

```
>> list := [a, b, c]: append(list, 3, 4, 5);
```

```
   [a, b, c, 3, 4, 5]
```

The dot operator . concatenates lists:

```
>> list1 := [1,2,3]: list2 := [4,5,6]:
>> list1.list2, list2.list1;
```

```
   [1, 2, 3, 4, 5, 6], [4, 5, 6, 1, 2, 3]
```

The corresponding system function is _concat and accepts arbitrarily many arguments. You can use it to combine many lists:

```
>> _concat(list1 $ 5);
```

```
   [1, 2, 3, 1, 2, 3, 1, 2, 3, 1, 2, 3, 1, 2, 3]
```

A list can be sorted by means of the function sort. This arranges numerical values according to their magnitude, strings (Section 4.11) are sorted lexicographically:

```
>> sort([-1.23, 4, 3, 2, 1/2]);
```

```
   [-1.23, 1/2, 2, 3, 4]
```

```
>> sort(["A", "b", "a", "c", "C", "c", "B", "a1", "abc"]);
```

```
   ["A", "B", "C", "a", "a1", "abc", "b", "c", "c"]
```

Note that the lexicographical order only applies to strings generated with ". Names of identifiers are sorted according to different (internal) rules, which take the length of the names into account:

```
>> sort([A, b, a, c, C, c, B, a1, abc]);
```

```
   [A, B, C, a, b, c, c, a1, abc]
```

MuPAD regards lists of function names as list-valued functions:

```
>> ([sin, cos, f])(x);

   [sin(x), cos(x), f(x)]
```

Note that the list must be put in parenthesis. The function map applies a function to all elements of a list:

```
>> map([x, 1, 0, PI, 0.3], sin);

   [sin(x), sin(1), 0, 0, 0.2955202066]
```

If the function has more than one argument, then map substitutes the list elements for the first argument and takes the remaining arguments from its own argument list:

```
>> map([a, b, c], f, y, z);

   [f(a, y, z), f(b, y, z), f(c, y, z)]
```

This map construction is a powerful tool for handling lists as well as other MuPAD objects. In the following example, we have a nested list L. We want to extract the first elements of the sublists using op(..,1). This is easily done using map:

```
>> L := [[a1, b1], [a2, b2], [a3, b3]]: map(L, op, 1);

   [a1, a2, a3]
```

The MuPAD function select enables you to extract elements with a certain property from a list. To this end, you need a function that checks whether an object has this property and returns TRUE or FALSE. For example, the call has(Object,object) returns TRUE if object is an operand or suboperand of Object, and otherwise FALSE:

```
>> has(1 + sin(1 + x), x), has(1 + sin(1 + x), y);

   TRUE, FALSE
```

Now

```
>> select([a + 2, x, y, z, sin(a)], has, a);
```

$$[a + 2, \sin(a)]$$

extracts all those list elements for which has(.,a) returns TRUE, i.e., those which contain the identifier a.

The function split divides a list into three lists, as follows. The first list contains all elements with a certain property, the second list collects all those elements without the property. If the test for the property returns the value UNKNOWN for some elements, then these are put into the third list. Otherwise the third list is empty. The function split returns a list comprising the three lists described above:

```
>> split([sin(x), x^2, y, 11], has, x);
```

$$[[\sin(x), x^2], [y, 11], []]$$

The MuPAD function zip combines elements of two lists pairwise into a new list:

```
>> L1 := [a, b, c]: L2 := [d, e, f]:
>> zip(L1, L2, _plus), zip(L1, L2, _mult),
   zip(L1, L2, _power);
```

$$[a + d, b + e, c + f], [a\,d, b\,e, c\,f], [a^d, b^e, c^f]$$

The third argument of zip must be a function that takes two arguments. This function is then applied to the pairs of list elements. In the above example, we have used the MuPAD functions _plus,_mult, and _power for addition, multiplication, and exponentiation, respectively. If the two input lists have different lengths, then the behavior of zip depends on the optional fourth argument. If this is not present, then the length of the resulting list is the minimum of the lengths of the two input lists. Otherwise, if you supply an additional fourth argument, then zip replaces the "missing" list entries by this argument:

```
>> L1 := [a, b, c, 1, 2]: L2 := [d, e, f]:
>> zip(L1, L2, _plus);

   [a + d, b + e, c + f]

>> zip(L1, L2, _plus, hello);

   [a + d, b + e, c + f, hello + 1, hello + 2]
```

Here is a summary of all list operations that we have discussed:

. or _concat	:	concatenating lists
append	:	appending elements
contains(list,x)	:	does list contain the element x?
list[i]	:	accessing the ith element
map	:	applying a function
nops	:	length
op	:	accessing elements
select	:	select according to properties
sort	:	sorting
split	:	split according to properties
subsop	:	replacing elements
unassign	:	deleting elements
zip	:	combining two lists

Exercise 4.14: Generate two lists with the elements a, b, c, d and $1, 2, 3, 4$, respectively. Concatenate the lists. Multiply the lists pairwise.

Exercise 4.15: Multiply all entries of the list [1,x,2] by 2. Suppose you are given a list, whose elements are lists of numbers or expressions, such as [[1, x, 2], [PI], [2/3, 1]], how can you multiply all entries by 2?

Exercise 4.16: Let $X = [x_1, \ldots, x_n]$ and $Y = [y_1, \ldots, y_n]$ be two lists of the same length. Find a simple method to compute

- their "inner product" (X as row vector and Y as column vector)

$$x_1 y_1 + \cdots + x_n y_n,$$

- their "matrix product" (X as column vector and Y as row vector)

$$[[x_1 y_1, x_1 y_2, \dots, x_1 y_n], [x_2 y_1, x_2 y_2, \dots, x_2 y_n],$$
$$\dots, [x_n y_1, x_n y_2, \dots, x_n y_n]\,.]$$

You can achieve this by using `zip`, `_plus`, `map` and appropriate functions (Section 4.12) within a single command line in each case. Loops (Chapter 16) are not required.

Exercise 4.17: In number theory, one is often interested in the density of prime numbers in sequences of the form $f(1), f(2), \dots$, where f is a polynomial. For each value of $m = 0, 1, \dots, 41$, find out how many of the integers $n^2 + n + m$ with $n = 1, 2, \dots, 100$ are primes.

Exercise 4.18: In which ordering will n children be eliminated by a counting-out rhyme composed of m words? For example, using

"eenie–meenie–miney–moe–catch a–tiger–by the–toe",

12 children are eliminated in the order 8–4–1–11–10–12–3–7–6–2–9–5. Hint: represent the children by a list `[1,2,..]` and remove an element from this list after it is counted out.

4.7 Sets

A *set* is an unordered sequence of arbitrary objects enclosed in curly braces. Sets are of domain type `DOM_SET`:

```
>> {34, 1, 89, x, -9, 8};
```

```
   {89, x, 1, 8, -9, 34}
```

The order of the elements in a MuPAD list appears to be random. The MuPAD kernel sorts the elements according to internal principles. You should use sets only if the order of the elements does not matter. If you want to process a sequence of expressions in a certain order, use lists, as discussed in the previous section.

Sets may be empty:

```
>> emptyset := {};
```

```
   {}
```

A set contains each element only once, i.e., duplicate elements are removed automatically:

```
>> set := {1, 2, 3, 4, a, b, 1, 2, a};

    {a, b, 1, 2, 3, 4}
```

The function **nops** determines the number of elements in a list. As for sequences and lists, **op** extracts elements from a set:

```
>> op(set);

    a, b, 1, 2, 3, 4

>> op(set, 2..4);

    b, 1, 2
```

Warning: Since elements of a set may be reordered internally, you should check carefully whether it makes sense to access the ith element. For example, **subsop(set,i=newvalue)** (Section 6) replaces the ith element by a new value. However, you should check in advance that the element that you want to replace really is the ith element.

Warning: In MuPAD versions beyond 1.4, there is a distinction between the order in which the elements of a set are stored internally and the order in which they are printed on the screen[15]. The command **op(set,i)** returns the ith element of **set** in the internal order, which usually is different from the ith element of **set** that you see on the screen. However, in versions beyond 1.4 you can access elements by using **set[i]**, where the returned elements is *guaranteed* to be the ith element as printed on the screen.

The functions **union**, **intersect**, and **minus** form the union, the intersection, and the set-theoretic difference, respectively, of sets:

```
>> M1 := {1, 2, 3, a, b}: M2 := {a, b, c, 4, 5}:
```

---

[15] The main reason for this change is the following. It can happen in versions up to 1.4 that two equivalent sets, such as, for example, {2, 3, 4} and {4, 3, 2}, are printed differently on the screen. In versions beyond 1.4, however, it is guaranteed that two sets whose elements agree up to reordering are printed on the screen in identical order.

```
>> M1 union M2, M1 intersect M2, M1 minus M2, M2 minus M1;
```

{a, b, c, 1, 2, 3, 4, 5}, {a, b}, {1, 2, 3}, {c, 4, 5}

In particular, you can use **minus** to remove elements from a set:

```
>> {1, 2, 3, a, b} minus {3, a};
```

{b, 1, 2}

You can also replace an element by a new value, without caring about the order of the elements:

```
>> set := {a, b, oldvalue, c, d}:
>> set minus {oldvalue} union {newvalue};
```

{a, newvalue, b, c, d}

The function **contains** checks whether an element belongs to a set, and returns either TRUE or FALSE[16]:

```
>> contains({a, b, c}, a), contains({a, b, c + d}, c);
```

TRUE, FALSE

MuPAD considers sets of function names as set-valued functions:

```
>> ({sin, cos, f})(x);
```

{cos(x), sin(x), f(x)}

Note that the set must be put in curly braces. You can apply a function to all elements of a set by means of **map**:

```
>> map({x, 1, 0,  PI, 0.3}, sin);
```

{0, 0.2955202066, sin(x), sin(1)}

---

[16] Note the difference to the behavior of **contains** for lists: there the ordering of the elements is determined when you generate the list, and **contains** returns the position of the element in the list.

You can use the function **select** to extract elements with a certain property from a set. This works as for lists, but the returned object is a set:

```
>> select({{a, x, b}, {a}, {x, 1}}, contains, x);

   {{x, 1}, {x, a, b}}
```

Similarly, you can use the function **split** to divide a set into three subsets of elements with a certain property, elements without that property, and elements for which the system cannot decide this and returns UNKNOWN. The result is a list comprising these three sets:

```
>> split({{a, x, b}, {a}, {x, 1}}, contains, x);

   [{{x, 1}, {x, a, b}}, {{a}}, {}]
```

Here is a summary of the set operations discussed so far:

contains(M, x)	:	does M contain the element x?
intersect	:	intersection
map	:	applying a function
minus	:	set-theoretic difference
nops	:	number of elements
op	:	accessing elements
select	:	select according to properties
split	:	split according to properties
subsop	:	replacing elements
union	:	set-theoretic union

The **combinat** library contains some combinatorial functions for finite sets. You may call **?combinat** to obtain a survey. For example, this package contains the function **combinat::powerset**, which returns the power set of a given set (see **?combinat::powerset** for more information).

MuPAD also provides the data structure **Dom::DiscreteSet** for handling (countably) finite sets. If $f(n)$ is an expression depending on n and a, b are integers, then **Dom::DiscreteSet(f(n),n=a..b)** represents the set

$$\{f(a), f(a+1), .., f(b)\}.$$

Both a and b may also assume the values -infinity or infinity. Thus Dom::DiscreteSet(n,n=-infinity..infinity) is a MuPAD object representing the set of integers.

Exercise 4.19: How can you convert a list to a set and vice versa?

Exercise 4.20: Generate the sets $A = \{a, b, c\}$, $B = \{b, c, d\}$, and $C = \{b, c, e\}$. Compute the union and the intersection of the three sets, as well as the difference $A \setminus (B \cup C)$.

Exercise 4.21: Instead of the binary operators intersect and union, you may also use the corresponding MuPAD functions _intersect and _union to compute unions and intersections of sets. These functions accept arbitrarily many arguments. Use simple commands to compute the union and the intersection of all sets belonging to M:

```
>> M := {{2, 3}, {3, 4}, {3, 7}, {5, 3}, {1, 2, 3, 4}}:
```

Exercise 4.22: The combinat library contains a function for generating all subsets of cardinality k of a finite set. Find this function and read the corresponding help page. Generate all subsets of $\{5, 6, \ldots, 20\}$ with 3 elements. How many of them are there?

4.8 Tables

A table is a MuPAD object of domain type DOM_TABLE, and represents a collection of equations of the form index=value. Both indices and values may be arbitrary MuPAD objects. You can generate a table by using the system function table ("explicit generation"):

```
>> T := table(a = b, c = d);

   table(
     a = b,
     c = d
   )
```

You can generate more entries or change existing ones by "indexed assignments" of the form Table[index]:=value:

```
>> T[f(x)] := sin(x): T[1, 2] := 5:
>> T[1, 2, 3] := {a, b, c}: T[a] := B:
>> T;

   table(
      (1, 2, 3) = {a, b, c},
      a = B,
      c = d,
      f(x) = sin(x),
      (1, 2) = 5
   )
```

It is not necessary to initialize a table via **table**. If T is an identifier that does not have a value, then an indexed assignment of the form T[index]:=value automatically turns T into a table ("implicit generation"):

```
>> unassign(T): T[a] := b: T[b] := c: T;

   table(
      a = b,
      b = c
   )
```

A table may be empty:

```
>> T := table();

   table()
```

You may delete table entries via **unassign(Table[index])**[17]:

```
>> T := table(a = b, c = d, d = a*c):
>> unassign(T[a], T[c]): T;

   table(
      d = a c
   )
```

---

[17] The keyword **delete** replaces the function **unassign** in MuPAD versions beyond 1.4.

You can access table entries in the form `Table[index]`; this returns the element corresponding to the index. If there is no entry for the index, then MuPAD returns `Table[index]` symbolically:

```
>> T := table(a = b, c = d, d = a*c):
>> T[a], T[b], T[c], T[d];

    b, T[b], d, a c
```

The call `op(Table)` returns all entries of a table, i.e., the sequence of all equations `index=value`:

```
>> op(table(a = A, b = B, c = C, d = D));

    a = A, b = B, c = C, d = D
```

Note that the internal order of the table entries may differ from the order in which you have generated the table. It may look quite random:

```
>> op(table(a.i = i^2 $ i = 1..10));

    a1 = 1, a2 = 4, a10 = 100, a3 = 9, a4 = 16, a5 = 25,

        a6 = 36, a7 = 49, a8 = 64, a9 = 81
```

The function `map` applies a given function to the *values* (not the *indices*) of all table entries:

```
>> T := table(1 = PI, 2 = 4, 3 = exp(1)): map(T, float);

    table(
      1 = 3.141592653,
      2 = 4.0,
      3 = 2.718281828
    )
```

The function `contains` checks whether a table contains a particular *index*. It ignores the *values*:

```
>> T := table(a = b): contains(T, a) , contains(T, b);

   TRUE, FALSE
```

You may employ the functions **select** and **split** to inspect both indices and values of a table and to extract them according to certain properties. This works similarly as for lists (Section 4.6) and for sets (Section 4.7):

```
>> T := table(1 = "number", 1.0 = "number", x = "symbol"):
>> select(T, has, "symbol");

   table(
     x = "symbol"
   )

>> select(T, has, 1.0);

   table(
     1.0 = "number"
   )

>> split(T, has, "number");

   -- table(              table(                    --
   |    1 = "number",  ,    x = "symbol" , table() |
   |    1.0 = "number"    )                        |
   -- )                                            --
```

Tables are particularly well suited for storing large amounts of data. Indexed accesses to *individual* elements are implemented efficiently also for big tables: a write or read does not file through the whole data structure.

Exercise 4.23: Generate a table **telephoneDirectory** with the following entries:

Ford 1815 , Reagan 4711 , Bush 1234 , Clinton 5678 .

Look up Ford's number. How can you find out whose number is 5678?

Exercise 4.24: Given a table, how can you generate a list of all indices and a list of all values, respectively?

Exercise 4.25: Generate the table `table(1=1,2=2,..,n=n)` and the list `[1,2,..,n]` of length $n = 100000$. Add a new entry to the table and to the list. How much time does this take? Hint: the call `time((a:=b))` returns the execution time for an assignment.

4.9 Arrays

Arrays, of domain type `DOM_ARRAY`, may be regarded as special tables. You may think of them as a collection of equations of the form `index=value`, but in contrast to tables, the indices must be non-negative integers. A one-dimensional array consists of equations of the form `i=value`. Mathematically, it represents a vector whose ith component is `value`. A two-dimensional array represents a matrix, whose (i,j)th component is stored in the form `(i,j)=value`. You may generate arrays of arbitrary dimension, with entries of the form `(i,j,k,..)=value`.

The system function `array` generates arrays. In its simplest form, you only specify a sequence of ranges that determine the dimension and the size of the array:

```
>> A := array(0..1, 1..3);
```

```
+-                             -+
|  ?[0, 1], ?[0, 2], ?[0, 3]  |
|                             |
|  ?[1, 1], ?[1, 2], ?[1, 3]  |
+-                             -+
```

You can see here that the first range `0..1` and the second range `1..3` determine the array's number of rows and columns, respectively. The output `?[0,1]` signalizes that the corresponding index has not been assigned a value yet. Thus the above command has generated an empty array. Now you can assign values to the indices:

```
>> A[0, 1] := 1: A[0, 2] := 2: A[0, 3] := 3:
>> A[1, 3] := HELLO: A;
```

```
+-                      -+
|     1,        2,      3    |
|                            |
|   ?[1, 1], ?[1, 2], HELLO  |
+-                      -+
```

You can also initialize the complete array directly when generating it
by **array**. Just supply the values as a (nested) list:

```
>> A := array(1..2, 1..3, [[1, 2, 3], [4, 5, 6]]);
```

```
+-        -+
|  1, 2, 3  |
|           |
|  4, 5, 6  |
+-        -+
```

You can access and modify array elements in the same way as table
elements:

```
>> A[2, 3] := A[2, 3] + 10: A;
```

```
+-         -+
|  1, 2,  3  |
|            |
|  4, 5, 16  |
+-         -+
```

Again, you may delete an array element by using **unassign**:

```
>> unassign(A[1, 1], A[2, 3]):  A , A[2, 3];
```

```
+-                 -+
|  ?[1, 1], 2,    3  |
|                     |, A[2, 3]
|      4,    5, ?[2, 3] |
+-                 -+
```

Arrays possess a "0th operand" op(**Array**,0), which provides information on the dimension and the size of the array. The call op(**Array**,0)

returns a sequence $d, a_1..b_1, \ldots, a_d..b_d$, where d is the dimension (i.e., the number of indices) and $a_i..b_i$ is the valid range for the ith index:

```
>> Vector := array(1..3, [x, y, z]): op(Vector, 0);
```

```
     1, 1..3
```

```
>> Matrix := array(1..2, 1..3, [[a, b, c], [d, e, f]]):
>> op(Matrix, 0);
```

```
     2, 1..2, 1..3
```

Thus the dimension of an $m \times n$ matrix `array(1..m,1..n)` may be obtained as

```
     m = op(Matrix,[0,2,2]), n = op(Matrix,[0,3,2]).
```

The internal structure of arrays differs from the structure of tables. The entries are not stored in the form of equations:

```
>> op(Matrix);
```

```
     a, b, c, d, e, f
```

The table type is more flexible than the array type: tables admit arbitrary indices, and their size may grow dynamically. Arrays are intended for storing vectors and matrices of a fixed size. When you enter an indexed call, the system checks whether the indices are within the specified ranges. For example:

```
>> Matrix[4, 7];
```

```
     Error: Illegal argument [array]
```

You may apply a function to all array components via `map`. For example, here is the simplest way to convert all array entries to floating-point numbers:

```
>> A := array(1..2, [ PI, 1/7]): map(A, float);
```

```
+-                          -+
| 3.141592653, 0.1428571428 |
+-                          -+
```

Warning: If m is an identifier without a value, then an indexed assignment of the form M[index,index,..]:=value generates a table and not an array of type DOM_ARRAY (Section 4.8):

```
>> unassign(M): M[1, 1] := a: M;

    table(
      (1, 1) = a
    )
```

Additionally, MuPAD provides the more powerful data structures of domain type Dom::Matrix for handling vectors and matrices. These are discussed in Section 4.15. Such objects are very convenient to use: you can multiply two matrices or a matrix and a vector by means of the usual multiplication symbol *. Similarly, you can add matrices of equal dimension via +. To achieve the same functionality with arrays, you have to write your own procedures. We refer to the examples MatrixProduct and MatrixMult in Sections 18.4 and 18.5, respectively.

Exercise 4.26: Generate what is called a Hilbert matrix, of dimension 20×20 with entries $H_{ij} = 1/(i + j - 1)$ for $i, j \geq 1$.

4.10 Boolean Expressions

MuPAD implements three logical ("Boolean") values: TRUE, FALSE, and UNKNOWN:

```
>> domtype(TRUE), domtype(FALSE), domtype(UNKNOWN);

    DOM_BOOL, DOM_BOOL, DOM_BOOL
```

The operators **and**, **or**, and **not** operate on Boolean values:

```
>> TRUE and FALSE, not (TRUE or FALSE), TRUE and UNKNOWN,
   TRUE or UNKNOWN;
```

```
   FALSE, FALSE, UNKNOWN, TRUE
```

The function **bool** evaluates equations, inequalities, or comparisons via >, >=, < , <=, to TRUE or FALSE:

```
>> a := 1: b := 2:
>> bool(a = b), bool(a <> b),
   bool(a <= b) or not bool(a > b);
```

```
   FALSE, TRUE, TRUE
```

Typically, you will use these constructs in branching conditions of **if** instructions (Chapter 17) or in termination conditions of **repeat** loops (Chapter 16). In the following example, we test the integers $1, \ldots, 3$ for primality. The system function **isprime** ("is the argument a prime number?") returns TRUE or FALSE. The **repeat** loop stops as soon as the termination condition $i = 3$ evaluates to TRUE:

```
>> i := 0:
   repeat
     i := i + 1;
     if isprime(i)
        then print(i, "is a prime")
        else print(i, "is no prime")
     end_if
   until i = 3 end_repeat;
```

```
   1, "is no prime"

   2, "is a prime"

   3, "is a prime"
```

Here we have used strings enclosed in " for the screen output. They are discussed in detail in Section 4.11. Note that it is not necessary to use the function **bool** in branching or termination conditions in order to evaluate the condition to TRUE or FALSE.

Exercise 4.27: Let ∧ denote the logical "and", let ∨ denote the logical "or", let ¬ denote logical negation. To which Boolean value does

$$\text{TRUE} \wedge (\text{FALSE} \vee \neg (\text{FALSE} \vee \neg \text{FALSE}))$$

evaluate?

Exercise 4.28: Let L1, L2 be two MuPAD lists of equal length. How can you find out whether L1[i]<L2[i] holds true for all list elements?

4.11 Strings

Strings are pieces of text, which may be used to format the screen output. A string is a sequence of arbitrary symbols enclosed in "string delimiters" ". Its domain type is DOM_STRING.

```
>> string1 := "Use * for multiplication";
   string2 := ", ";
   string3 := "use ^ for exponentiation.";

   "Use * for multiplication"

   ", "

   "use ^ for exponentiation."
```

The concatenation operator . combines strings:

```
>> string4 := string1.string2.string3;

   "Use * for multiplication, use ^ for exponentiation."
```

The dot operator is a short form of the MuPAD function _concat, which concatenates (arbitrarily many) strings:

```
>> _concat("This is ", "a string", ".");

   "This is a string."
```

The index operator [ ] extracts the characters from a string:

```
>> string4[0], string4[1], string4[2], string4[3],
   string4[4];
```

   "U", "s", "e", " ", "*"

You may use the command **print** to output intermediate results in loops or procedures on the screen (Section 13.1.1). By default, this function prints strings with the enclosing double quotes. You may change this behavior by using the optional argument Unquoted:

```
>> print(string4);
```

   "Use * for multiplication, use ^ for exponentiation."

```
>> print(Unquoted, string4);
```

   Use * for multiplication, use ^ for exponentiation.

Strings are no valid identifiers in MuPAD, so you cannot assign values to them:

```
>> "name" := sin(x);
```

   Syntax Error: Unexpected symbol in assignment

   "name" := sin(x);
         ^

Also, arithmetic with strings is not allowed:

```
>> 1 + "x";
```

   Error: Illegal operand [_plus]

However, you may use strings in equations:

```
>> "derivative of sin(x)" = cos(x);
```

   "derivative of sin(x)" = cos(x)

The function `expr2text` converts a MuPAD object to a string. You can employ this function to customize print commands:

```
>> i := 7:
>> print(Unquoted, expr2text(i)." is a prime.");

   7 is a prime.

>> a := sin(x):
>> print(Unquoted, "The derivative of " . expr2text(a) .
                   " is " . expr2text(diff(a, x)). ".");

   The derivative of sin(x) is cos(x).
```

You find numerous other useful functions for handling strings in the standard library (Section "Manipulation of Strings" of the MuPAD Quick Reference [Oev 98]) and in the string library (`?string`)[18].

Exercise 4.29: In Section 4.3, we have already mentioned the command `anames(3)`, which returns a list of all identifiers that have a value in the current session. Generate a *lexicographically* ordered list of these identifiers.

Exercise 4.30: How can you obtain the "mirror image" of a string? Hint: the function `strlen` (short for: string length) returns the number of symbols in a string. In MuPAD versions beyond 1.4, this function is renamed to `length`.

4.12 Functions

The arrow operator `->`, built from a minus sign and a "greater than" sign, generates simple objects that represent mathematical functions[19]:

```
>> f := (x, y) -> (x^2 + y^2);

   (x, y) -> (x^2 + y^2)
```

---

[18] The library `string` is renamed to `stringlib` in MuPAD versions beyond 1.4.

[19] In MuPAD versions up to 1.4, this generates a separate data structure of domain type `DOM_EXEC`: executable function. In later versions, there is no longer a difference between functions and procedures (Chapter 18). Objects generated via `->` are then of domain type `DOM_PROC`.

The function f can now be called like any system function. It takes two arbitrary input parameters (or "arguments") and returns the sum of their squares:

```
>> f(a, b + 1);

    2         2
   a  + (b + 1)
```

In the following example, the return value of the function is generated by an if statement:

```
>> absValue := x -> (if x >= 0 then x else -x end_if):
>> absValue(-2.3);

   2.3
```

Warning: In MuPAD versions up to 1.4, the system evaluates all formal parameters and identifiers when you define a function via ->:

```
>> y := 1: f := x -> (x + y);

   x -> (x + 1)
```

For that reason, the formal parameters should have no value:

```
>> x := 1: f := x -> (x + z);

   1 -> (z + 1)

>> unassign(x): f := x -> (x + z);

   x -> (x + z)
```

Function names are also evaluated and the corresponding algorithm is executed. For example, suppose that we want to define a function returning the first element of a list by using op(list,1):

```
>> f := list -> (op(list, 1));

   list -> list
```

Here, the call of op returns the first (and only) operand of its input, the symbolic identifier list, and hence f is the identity function. To achieve what we wanted, we delay the evaluation by means of hold (Chapter 5.2):

```
>> f := list -> (hold(op)(list, 1)); f([sin(x), b, c]);

  list -> (op(list, 1))


  sin(x)
```

This behavior changes in MuPAD versions beyond 1.4. Then the system evaluates *neither the formal parameters nor identifiers* in a function definition, and f := list -> op(list,1) indeed generates a function returning the first list element.

As discussed in Section 4.4.1, the operator @ generates the composition $h : x \to f(g(x))$ of two functions f and g:

```
>> f := x -> (1/(1 + x)): g := x -> (sin(x^2)):
>> h := f@g: h(a);

         1
    -----------
          2
    sin(a ) + 1
```

You can define a repeated composition $f(f(f(.)))$ of a function with itself by using the iteration operator @@:

```
>> fff := f@@3: fff(a);

           1
      -------------
           1
      --------- + 1
        1
      ----- + 1
      a + 1
```

Of course, these constructions also work for system functions. For example, the function abs@Re computes the absolute value of the real part of a complex number:

```
>> f := abs@Re: f(-2 + 3*I);
```

$$2$$

In symbolic computations, you often have the choice to represent a mathematical function either as a *map* **arguments** $\rightarrow$ **value** or as an *expression*:

```
>> Map := x -> (2*x*cos(x^2)):
>> Expression := 2*x*cos(x^2):
>> int(Map(x), x), int(Expression, x);
```

$$\text{sin}(x^2), \text{sin}(x^2)$$

You can easily convert between these two kinds of representation. For example, the function unapply from the fp library converts an expression to a function:

```
>> h := fp::unapply(Expression);
```

$$2 \text{ id } \text{cos@}(\text{id}^2)$$

Here, id denotes the identity map[20]. Indeed h and Map represent the same function:

```
>> h(x) = Map(x);
```

$$2 x \text{ cos}(x^2) = 2 x \text{ cos}(x^2)$$

---

[20] In MuPAD version 1.4, h is a functional expression, of domain type DOM_EXPR. In later MuPAD versions, fp::unapply(Expression) returns a function of domain type DOM_PROC, namely x -> 2*x*cos(x^2).

The function **h** is an example for MuPAD's feature to represent maps by means of *functional expressions*: more complex functions are constructed from simple functions (such as **sin**, **cos**, **exp**, **ln**, **id**) by means of operators (such as the composition operator **@** or the arithmetic operators **+**, *****, etc.). Note that the arithmetic operators generate functions that are defined *pointwise*, which is mathematically sound. For example, $f + g$ represents the map $x \to f(x) + g(x)$, $f\,g$ represents the map $x \to f(x)\,g(x)$, etc.:

```
>> unassign(f, g):
>> a := f + g: b := f*g: c := f/g: a(x), b(x), c(x);
```

$$
f(x) + g(x), \quad f(x)\,g(x), \quad \frac{f(x)}{g(x)}
$$

You are allowed to have numerical values in functional expressions. MuPAD regards them as constant functions which always return the particular value:

```
>> a := f + 1: b := f*3/4: c := f + 0.1: d := f + sqrt(2):
>> a(x), b(x), c(x), d(x);
```

$$
f(x) + 1, \quad \frac{3\,f(x)}{4}, \quad f(x) + 0.1, \quad f(x) + 2^{1/2}
$$

The operator **->** is useful for defining functions whose return value can be obtained by simple operations. Functions implementing more complex algorithms usually require many commands and auxiliary variables to store intermediate results. In principle, you can define such functions via **->** as well. However, this has the drawback that you often use *global variables*. Instead, we recommend to define a procedure via **proc() begin .. end_proc**. This concept of MuPAD's programming language is much more flexible and is discussed in more detail in Chapter 18.

Exercise 4.31: Define the functions $f(x) = x^2$ and $g(x) = \sqrt{x}$. Compute $f(f(g(2))$ and $\underbrace{f(f(\ldots f(x)..))}_{100 \text{ times}}$.

Exercise 4.32: Define a function that reverses the order of the elements in a list.

Exercise 4.33: The *Chebyshev polynomials* are defined recursively by the following formulae:

$$T_0(x) = 1 , \quad T_1(x) = x , \quad T_k(x) = 2\,x\,T_{k-1}(x) \ - \ T_{k-2}(x) .$$

Compute the values of $T_2(x), \ldots, T_5(x)$ for $x = 1/3$, $x = 0.33$, and for a symbolical value x.

4.13 Series Expansions

Expressions such as `1/(1-x)` admit series expansions for symbolic parameters. This particularly simple example is the sum of the geometric series:

$$\frac{1}{1 - x} = 1 + x + x^2 + x^3 + \cdots .$$

The function `taylor` computes the leading terms of such series:

```
>> t := taylor(1/(1 - x), x = 0, 9);
```

$$1 + x + x^2 + x^3 + x^4 + x^5 + x^6 + x^7 + x^8 + O(x^9)$$

This is the Taylor series expansion of the expression around the point $x = 0$, as requested by the second argument. MuPAD has truncated the infinite series before the term x^9 and has collected the tail in the "big Oh" term $O(x^9)$. The (optional) third argument of `taylor` controls the truncation. If it is not present, then MuPAD substitutes the value of the environment variable `ORDER` instead, whose default value is 6:

```
>> t := taylor(1/(1 - x), x = 0);
```

$$1 + x + x^2 + x^3 + x^4 + x^5 + O(x^6)$$

The resulting series looks like an ordinary sum with an additional `O(.)` term. Internally, however, it is represented by a special data structure of domain type `Puiseux`[21]:

---

[21] This type is called `Series::Puiseux` in MuPAD versions beyond 1.4.

```
>> domtype(t);
```

    Puiseux

The big-Oh term itself is a data structure on its own, of domain type
O and with special rules of manipulation:

```
>> 2*O(x^2) + O(x^3), x^2*O(x^10), O(x^5)*O(x^20),
   diff(O(x^3), x);
```

$$O(x^2), \; O(x^{12}), \; O(x^{25}), \; O(x^2)$$

The ordering of the terms in a Taylor series is fixed: powers with
smaller exponents precede those with higher exponents. This is in contrast to the ordering in ordinary sums, which looks quite random:

```
>> S := expr(t);
```

$$x + x^2 + x^3 + x^4 + x^5 + 1$$

Here we have used the system function **expr** to convert the series to
an expression of domain type **DOM_EXPR**. As you can see in the output,
the O(.) term has been cut off.

The **op** command acts on series in a non-obvious way:

```
>> op(t);
```

    1, 0, 6, [1, 1, 1, 1, 1, 1], x = 0

The first operand is the "branching degree" and provides information
about the nonuniqueness of the expansion[22]. The second and the third
operand denote the smallest exponent in the expansion and the $O()$
term, respectively. The fourth operand is a list with the coefficients.
The last operand is the point of expansion.

Alternatively, you may use the function **coeff** to extract the coefficients. This is more intuitive than **op**. The call **coeff(t,i)** returns
the coefficient of x^i:

---

[22] This is relevant if you want to expand multi-valued functions, such as $\sqrt{x}$ around
$x = 0$. This is accomplished by the function **series** rather than by **taylor**. The
latter also calls **series** internally.

```
>> t := taylor(cos(x^2), x, 20);
```

$$1 - \frac{x^4}{2} + \frac{x^8}{24} - \frac{x^{12}}{720} + \frac{x^{16}}{40320} + O(x^{20})$$

```
>> coeff(t, 0), coeff(t, 1), coeff(t, 4), coeff(t, 20);
```

1, 0, -1/2, FAIL

In the previous example, we have supplied x as second argument to specify the point of expansion. This is equivalent to x=0.

The usual arithmetic operations also work for series:

```
>> a := taylor(cos(x), x, 3): b := taylor(sin(x), x, 4):
>> a, b;
```

$$1 - \frac{x^2}{2} + O(x^3), \quad x - \frac{x^3}{6} + O(x^4)$$

```
>> a + b, 2*a*b, a^2;
```

$$1 + x - \frac{x^2}{2} + O(x^3), \quad 2x - \frac{4x^3}{3} + O(x^4), \quad 1 - x^2 + O(x^3)$$

Both the composition operator @ and the iteration operator @@ apply to series as well:[23]

```
>> a := taylor(sin(x), x, 20):
>> b := taylor(asin(x), x, 20): a@b;
```

$$x + O(x^{20})$$

---

[23] The function asin is renamed to arcsin in MuPAD versions beyond 1.4.

If you try to compute the Taylor series of a function that does not have
one, then `taylor` returns FAIL[24]. The function `series` can compute
more general expansions (Laurent series, Puiseux series):

```
>> taylor(cos(x)/x, x = 0, 10);

   FAIL
```

```
>> series(cos(x)/x, x = 0, 10);
```

$$\frac{1}{x} - \frac{x}{2} + \frac{x^3}{24} - \frac{x^5}{720} + \frac{x^7}{40320} + O(x^9)$$

You can generate series expansions in terms of negative powers by
expanding around the point `infinity`:

```
>> series((x^2 + 1)/(x + 1), x = infinity);
```

$$x - 1 + \frac{2}{x} - \frac{2}{x^2} + \frac{2}{x^3} + O\left(\frac{1}{x^4}\right)$$

This is an example for an "asymptotic" expansion, which approximates
the behavior of a function for large values of the argument. In simple
cases, `series` returns an expansion in terms of negative powers of x,
but other functions may turn up as well:

```
>> series((exp(x) - exp(-x))/(exp(x) + exp(-x)),
          x = infinity);
```

$$1 - \frac{2}{\exp(x)^2} + \frac{2}{\exp(x)^4} + O\left(\frac{1}{\exp(x)^5}\right)$$

---

[24] In MuPAD versions beyond 1.4, the computation is aborted with an error message.

Exercise 4.34: The order p of a root x of a function f is the maximal number of derivatives that vanish at the point x:

$$f(x) = f'(x) = \ldots = f^{(p-1)}(x) = 0 , \quad f^{(p)}(x) \neq 0 .$$

What is the order of the root $x = 0$ of $f(x) = \tan(\sin(x)) - \sin(\tan(x))$?

Exercise 4.35: Besides the arithmetical operators, some other system functions such as `diff` or `int` work directly for series. Compare the result of `taylor(cos(x),x)` and the derivative of `taylor(sin(x),x)`. Mathematically, both series are identical. Can you explain the difference in MuPAD?

Exercise 4.36: The function $f(x) = \sqrt{x+1} - \sqrt{x-1}$ has the limit $\lim_{x \to \infty} f(x) = 0$. Show that the approximation $f(x) \approx 1/\sqrt{x}$ is valid for large values of x. Find better asymptotic approximations of f.

Exercise 4.37: Compute the first three terms in the series expansion of the function `f:=sin(x+x^3)` around `x=0`. Read the help page for the MuPAD function `revert`. Use this function to compute the leading terms of the series expansion of the inverse function f^{-1} (which is well-defined in a certain neighborhood of $x = 0$).

4.14 Algebraic Structures: Fields, Rings, etc.

The MuPAD kernel provides domain types for the basic data structures such as numbers, sets, tables, etc. In addition, you can define your own data structures in the MuPAD language and work with them symbolically. We do not discuss the construction of such new "domains" in this elementary introduction, but demonstrate some special "library" domains provided by the system.

Besides the kernel domains, the Dom library contains a variety of predefined domains that were implemented by the MuPAD developers. The following command prints an overview:

```
>> info(Dom);
```

```
Library 'Dom': Basic domain constructors
Interface:
Dom::AlgebraicExtension,
Dom::ArithmeticalExpression,
Dom::BaseDomain,
Dom::Complex,
...
```

The help pages, such as ?Dom::Complex, provide information on individual data structures. In this section, we present some particularly useful domains representing complex mathematical objects such as fields, rings, etc. Section 4.15 discusses a data type for matrices, which is well-suited for problems in linear algebra.

The main part of a domain is its *constructor*, which generates objects of the domain. Each such object has *methods* attached to it. They represent the mathematical operations for these objects.

Here is a list of some of the well-known mathematical structures implemented in the Dom library:

- the ring of integers $\mathbb{Z}$: Dom::Integer,
- the field of rational numbers $\mathbb{Q}$: Dom::Rational,
- the field of real numbers $\mathbb{R}$: Dom::Real or Dom::Float[25],
- the field of complex numbers $\mathbb{C}$: Dom::Complex,
- the ring of integers modulo n: Dom::IntegerMod(n).

We consider the residue class ring of integers modulo n. Its elements are the integers $0, 1, \ldots, n-1$, and addition and multiplication are defined "modulo n". This works by adding or multiplying in $\mathbb{Z}$, dividing the result by n, and taking the remainder in $\{0, 1, \ldots, n-1\}$ of this division:

```
>> 3*5 mod 7;
```

```
                              1
```

In this example, we have used the data types of the MuPAD kernel: the operator * multiplies the integers 3 and 5 in the usual way to get 15, and the operator mod computes the decomposition $15 = 2 \cdot 7 + 1$ and returns 1 as remainder modulo 7.

---

[25] Dom::Real is for symbolic representations of real numbers, while Dom::Float represents them as floating-point numbers.

Alternatively, you may perform this computation in the domain Dom::IntegerMod(7). The latter object acts as a constructor for elements of the residue class ring[26] modulo 7:

```
>> constructor := Dom::IntegerMod(7):
>> x := constructor(3); y := constructor(5);
```

```
   3 mod 7
```

```
   5 mod 7
```

As you can see from the screen output, the identifiers x and y do not have the integers 3 and 5, respectively, as values. Instead, the numbers are internally marked as elements of the residue class ring of integers modulo 7:

```
>> domtype(x), domtype(y);
```

```
   Dom::IntegerMod(7), Dom::IntegerMod(7)
```

Now you can use the usual arithmetic operations, and MuPAD automatically uses the computation rules of the residue class ring:

```
>> x*y, x^123*y^17 - x + y;
```

```
   1 mod 7, 6 mod 7
```

The ring Dom::IntegerMod(7) even has a field structure, so that you can divide by all ring elements except 0 mod 7:

```
>> x/y;
```

```
   2 mod 7
```

---

[26] If you want to execute only a small number of modulo operations, then it is often preferable to use the operator mod, which is implemented in the MuPAD kernel and therefore quite fast. This approach may require some additional understanding how the system functions work. For example, the computation of 17^29999 mod 7 takes quite a long time, since MuPAD first computes the very big number 27^{29999} and then reduces the result modulo 7. In this case, the computation x^29999, where x:=Dom::IntegerMod(7)(17), is much faster, since the internal modular arithmetic avoids such big numbers. Alternatively, the call powermod(17,29999,7) uses the system function powermod to compute the result quickly without employing Dom::IntegerMod(7).

A more abstract example is the field extension

$$K = \mathbb{Q}[\sqrt{2}] = \{p + q\,\sqrt{2} \; ; \; p, q \in \mathbb{Q}\} \; .$$

You may define this field in MuPAD via

```
>> K := Dom::AlgebraicExtension(Dom::Rational,
                            Sqrt2^2 = 2, Sqrt2):
```

Here the identifier Sqrt2 ($\hat{=}\sqrt{2}$), defined by its algebraic property Sqrt2^2=2, is used to extend the rational numbers Dom::Rational. Now you can compute in this field:

```
>> x := K(1/2 + 2*Sqrt2): y := K(1 + 2/3*Sqrt2):
>> x^2*y + y^4;
```

```
  677 Sqrt2
  --------- + 5845/324
     54
```

The domain Dom::ExpressionField(normalizer,zerotest) represents the field of (symbolic) MuPAD expressions. The constructor has two functions **normalizer** and **zerotest**, which may be chosen by the user, as parameters.

The function **zerotest** is called internally by all algorithms that want to decide whether a domain object is mathematically 0. Typically, you will use the system function **iszero**, which recognizes not only the integer 0 as zero, but also other objects such as the floating-point number 0.0 or the trivial polynomial poly(0,[x]) (Section 4.16).

The task of the function **normalizer** is to generate a normal form for MuPAD objects of type Dom::ExpressionField(.,.). Operations on such objects will use this function to simplify the result before returning it. For example, if you supply the identity function **id** for the **normalizer** argument, then operations on objects of this domain work like for the usual MuPAD expressions, without additional normalization.

```
>> constructor := Dom::ExpressionField(id, iszero):
>> x := constructor(a/(a + b)^2):
   y := constructor(b/(a + b)^2):
```

```
>> x + y;
```

```
    a              b
-------- + --------
        2          2
 (a + b)    (a + b)
```

If you supply the system function **normal** instead, then the result is simplified automatically (Section 9.1):

```
>> constructor := Dom::ExpressionField(normal, iszero):
>> x := constructor(a/(a + b)^2):
   y := constructor(b/(a + b)^2):
>> x + y;
```

```
   1
 -----
 a + b
```

We note that the purpose of such MuPAD domains is not necessarily the direct generation of data structures or the computation with the corresponding objects. Indeed, some constructors simply return objects of the underlying kernel domains, if such domains exist:

```
>> domtype(Dom::Integer(2)),
   domtype(Dom::Rational(2/3)),
   domtype(Dom::Float(PI)),
   domtype(Dom::ExpressionField(id, iszero)(a + b));
```

```
DOM_INT, DOM_RAT, DOM_FLOAT, DOM_EXPR
```

In these cases, there is no immediate benefit in using such a constructor; you may as well compute directly with the corresponding kernel objects. The main application of such special data structures is for the construction of more complex mathematical structures. A simple example is the generation of matrices (Section 4.15) or polynomials (Section 4.16) with entries in a particular ring, such that matrix or polynomial arithmetic, respectively, is performed according to the computation rules of the coefficient ring.

4.15 Vectors and Matrices

In Section 4.14, we have given examples of special data types ("domains") for defining algebraic structures such as rings, fields, etc. in MuPAD. In this section, we discuss two further domains for generation and convenient computation with vectors and matrices: Dom::Matrix and Dom::SquareMatrix. In principle, you may use arrays for working with vectors or matrices (Section 4.9). However, then you have to define you own routines for addition, multiplication, inversion, or determinant computation, using MuPAD's programming language (Chapter 18). For the special matrix type that we present in what follows, such routines exist and are "attached" to the matrices as methods. Moreover, you may use the functions of the linalg library (linear algebra, Section 4.15.4), which can handle matrices of this type.

4.15.1 Definition of Matrices

MuPAD provides the two data types Dom::Matrix for rectangular matrices of arbitrary dimension $m \times n$ and Dom::SquareMatrix for square matrices of dimension $n \times n$. They belong to the library Dom, which also comprises data types for mathematical structures such as rings and fields (Section 4.14). Matrices may have entries from a set that must be equipped with a ring structure in the mathematical sense. For example, you may use the predefined rings and fields such as Dom::Integer, Dom::IntegerMod(n), etc. from the Dom library.

The function Dom::Matrix(R) is the constructor for rectangular matrices of arbitrary dimension $m \times n$ with coefficients in the ring R. When you construct such a matrix, you are required to ensure that its entries belong to (or may be converted to) this ring. You should keep this in mind when trying to generate matrices with entries outside the coefficient ring in a computation (for example, the inverse of an integer matrix in general has non-integral rational entries).

The following example yields the constructor for matrices with rational number entries[27]:

```
>> constructor := Dom::Matrix(Dom::Rational);

    Dom::Matrix(Dom::Rational)
```

---

[27] After exporting (Section 3.2) the Dom library via export(Dom), you may write constructor := Matrix(Rational) for short.

Now you may generate matrices of arbitrary dimensions. In the following example, we generate a 2×3 matrix with all entries initialized to 0:

```
>> matrix := constructor(2, 3);
```

```
+-          -+
|  0, 0, 0  |
|           |
|  0, 0, 0  |
+-          -+
```

When generating a matrix, you may supply a function f that takes two arguments. Then the entry in row i and column j is initialized with $f(i, j)$:

```
>> f := (i, j) -> (i*j):  matrix := constructor(2, 3, f);
```

```
+-          -+
|  1, 2, 3  |
|           |
|  2, 4, 6  |
+-          -+
```

Alternatively, you can initialize a matrix by specifying a (nested) list. Each list element is itself a list and corresponds to one row of the matrix. The following command generates the same matrix as in the previous example:

```
>> matrix := constructor(2, 3, [[1, 2, 3], [2, 4, 6]]):
```

The parameters for the dimension are optional here, since they are also given by the structure of the list. Thus

```
>> matrix := constructor([[1, 2, 3], [2, 4, 6]]):
```

also returns the same matrix. An array of domain type DOM_ARRAY (Section 4.9) is also valid for initializing a matrix:

```
>> Array := array(1..2, 1..3, [[1, 2, 3], [2, 4, 6]]):
>> matrix := constructor(Array):
```

You may define column and row vectors as $m \times 1$ and $1 \times n$ matrices, respectively. If you use lists for initializing them, then they need not be nested:

```
>> column := constructor(3, 1, [1, 2, 3]);
```

```
+-   -+
|  1  |
|     |
|  2  |
|     |
|  3  |
+-   -+
```

```
>> row := constructor(1, 3, [1, 2, 3]);
```

```
+-        -+
| 1, 2, 3 |
+-        -+
```

If you only supply a non-nested list, then this generates a column vector:

```
>> column := constructor([1, 2, 3]);
```

```
+-   -+
|  1  |
|     |
|  2  |
|     |
|  3  |
+-   -+
```

You may access the entries of a matrix in the form matrix[i,j], row[i], or column[j]. Since vectors are special matrices, you may access the components of a vector in the form row[1,i] or column[j,1], respectively, as well:

```
>> matrix[2, 3], row[3], row[1, 3],
   column[2], column[2, 1];

    6, 3, 3, 2, 2
```

Submatrices are generated as follows:

```
>> matrix[1..2, 1..2], row[1..1, 1..2],
   column[1..2, 1..1];

   +-        -+                +-     -+
   |  1, 2  |  +-      -+  |  1  |.
   |           |, | 1, 2 |, |     |
   |  2, 4  |  +-      -+  |  2  |
   +-        -+                +-     -+
```

You may change a matrix entry by an indexed assignment:

```
>> matrix[2, 3] := 23: row[2] := 5: column[2, 1] := 5:
>> matrix, row, column;
```

```
                                          +-    -+
                                          |  1  |
   +-            -+                    |      |
   |  1, 2,  3  |  +-          -+  |      |
   |               |, | 1, 5, 3 |, |  5  |
   |  2, 4, 23  |  +-          -+  |      |
   +-            -+                    |  3  |
                                          +-    -+
```

You can use loops (Chapter 16) to change all components of a matrix:

```
>> m := 2: n := 3: matrix := constructor(m, n):
>> for i from 1 to m do
     for j from 1 to n do
       matrix[i, j] := i*j
     end_for
   end_for:
```

You can generate diagonal matrices by supplying the option **Diagonal**.
In this case, the third argument to the constructor may either be a list

of the diagonal elements of a function f such that the ith diagonal element is $f(i)$:

```
>> constructor(2, 2, [11, 12], Diagonal);
```

```
   +-          -+
   |   11,  0  |
   |           |
   |    0, 12  |
   +-          -+
```

In the next example, we generate an identity matrix by supplying 1 as a function defining the diagonal elements[28]:

```
>> constructor(2, 2, 1, Diagonal);
```

```
   +-       -+
   |  1, 0  |
   |        |
   |  0, 1  |
   +-       -+
```

All constructors considered so far return matrices with rational (i.e., real) number entries. Thus the following attempt to generate a matrix with invalid complex coefficients does not work:

```
>> constructor([[1, 2, 3], [2, 4, 1 + I]]);
```

```
   Error: unable to define matrix over Dom::Rational
     [(Dom::Matrix(Dom::Rational))::new]
```

You have to choose a suitable coefficient ring to generate a matrix with the above entries. In the following example, we define a new constructor for matrices with complex number entries:

```
>> constructor := Dom::Matrix(Dom::Complex):
```

---

[28] MuPAD considers numbers as (constant) functions. The following example produces the same output, no matter what the value of x is:

```
   >> f := 1: f(x);
```

```
   1
```

```
>> constructor([[1, 2, 3], [2, 4, 1 + I]]);
```

```
+-            -+
|  1, 2,   3  |
|             |
|  2, 4, 1 + I |
+-            -+
```

You may generate matrices whose entries are arbitrary MuPAD expressions by means of the field Dom::ExpressionField(id,iszero) (see Section 4.14). This is the standard coefficient ring for matrices. You may always use this ring when the coefficients and their properties are irrelevant. Dom::Matrix with no argument is a constructor for such matrices:

```
>> constructor := Dom::Matrix();
```

```
Dom::Matrix(Dom::ExpressionField(id, iszero))
```

```
>> constructor([[1, x + y, 1/x^2], [sin(x), 0, cos(x)],
                [x*PI, 1 + I, -x*PI]]);
```

```
+-                         -+
|                     1     |
|      1,    x + y,   --     |
|                     2     |
|                     x     |
|                           |
|   sin(x),    0,    cos(x)  |
|                           |
|    x PI,  1 + I, - x PI    |
+-                         -+
```

If you use Dom::ExpressionField(normal,iszero) as coefficient ring, then all matrix entries are simplified via the function normal, as described in Section 4.14. Arithmetical operations with such matrices are comparatively slow, since a call to normal may be quite time consuming. On the other hand, the returned results are in general simpler than the (equivalent) results of computations that use Dom::ExpressionField(id,iszero).

The function Dom::SquareMatrix(n,R) generates the ring of n-dimensional square matrices with coefficient ring R. If the argument R is missing, then MuPAD automatically uses the coefficient ring of all MuPAD expressions. The following statement yields the constructor for square matrices with two rows, whose entries may be arbitrary MuPAD expressions:

```
>> constructor := Dom::SquareMatrix(2);

   Dom::SquareMatrix(2, Dom::ExpressionField(id, iszero))

>> constructor([[0, y], [x^2, 1]]);
```

```
+-         -+
|   0, y  |
|          |
|   2      |
|  x , 1  |
+-         -+
```

4.15.2 Computing with Matrices

You can use the standard arithmetical operators for doing basic arithmetic with matrices:

```
>> A := Dom::Matrix()([[1, 2], [3, 4]]):
>> B := Dom::Matrix()([[a, b], [c, d]]):
>> A + B, A*B, A*B - B*A, A^2 + B;
```

```
+-             -+ +-                        -+
|  a + 1, b + 2 | |  a + 2 c,    b + 2 d   |
|             |, |                        |,
|  c + 3, d + 4 | |  3 a + 4 c, 3 b + 4 d  |
+-             -+ +-                        -+

   +-                              -+
   |    - 3 b + 2 c ,  - 2 a - 3 b + 2 d  |
   |                               |,
   |  3 a + 3 c - 3 d,     3 b - 2 c  |
   +-                              -+
```

```
+-                    -+
|    a + 7, b + 10   |
|                    |
|   c + 15, d + 22   |
+-                    -+
```

Multiplication of a matrix and a number works componentwise (scalar multiplication):

```
>> 2*B;
```

```
+-           -+
|  2 a, 2 b  |
|            |
|  2 c, 2 d  |
+-           -+
```

The inverse of a matrix is represented by 1/A or A^(-1):

```
>> C := 1/A;
```

```
+-           -+
|   -2,   1  |
|            |
|  3/2, -1/2 |
+-           -+
```

A simple test shows that the computed inverse is correct:

```
>> A*C, C*A;
```

```
+-       -+ +-       -+
|  1, 0  | |  1, 0  |
|        |,|        |
|  0, 1  | |  0, 1  |
+-       -+ +-       -+
```

An inversion returns FAIL when MuPAD is unable to compute the result. The following matrix is not invertible:

```
>> C := Dom::Matrix()([[1, 1], [1, 1]]): C^(-1);
```

        FAIL

The concatenation operator ., which combines lists (Section 4.6) or strings (Section 4.11), is "overloaded" for matrices. You can use it to combine matrices with the same number of rows:

```
>> A, B, A.B;
```

```
+-        -+ +-        -+ +-                -+
|  1, 2  | |  a, b  | |  1, 2, a, b  |
|          |, |          |, |          |              |
|  3, 4  | |  c, d  | |  3, 4, c, d  |
+-        -+ +-        -+ +-                -+
```

Besides the arithmetic operators, other system functions are applicable to matrices. Here is a list of examples:

- `conjugate(A)` replaces all components by their complex conjugates,
- `diff(A,x)` differentiates componentwise with respect to x,
- `exp(A)` computes $e^A = \sum_{i=0}^{\infty} \frac{1}{i!} A^i$,
- `expand(A)` applies `expand` to all components of A,
- `expr(A)` converts A to an array of domain type `DOM_ARRAY`,
- `float(A)` applies `float` to all components of A,
- `has(A,expression)` checks whether an expression is contained in at least one entry of A,
- `int(A,x)` integrates componentwise with respect to x,
- `iszero(A)` checks whether all components of A vanish,
- `map(A,function)` applies the function to all components,
- `norm(A)` (identical with `norm(A,Infinity)`) computes the infinity norm[29],
- `subs(A,equation)` applies `subs(.,equation)` to all entries of A,

---

[29] `norm(A,1)` returns the one-norm, `norm(A,Frobenius)` the Frobenius norm $\left(\sum_{i,j} |A_{ij}|^2\right)^{1/2}$.

- C:=zip(A,B,f) returns the matrix defined by $C_{ij} = f(A_{ij}, B_{ij})$.

The linear algebra library linalg and the numerics library numeric (Section 4.15.4) comprise many other functions for handling matrices.

Exercise 4.38: Generate the 15×15 Hilbert matrix $H = (H_{ij})$ with $H_{ij} = 1/(i + j - 1)$. Generate the vector $\mathbf{b} = H\mathbf{e}$, where $\mathbf{e} = (1, \ldots, 1)$. Generate the exact solution vector $\mathbf{x}$ of the system of equations $H\mathbf{x} = \mathbf{b}$ (of course, this should yield $\mathbf{x} = \mathbf{e}$). Convert all entries of H to floating-point values and solve the system of equations again. Compare the result to the exact solution. You will note a dramatic difference, which originates from numerical rounding errors. Larger Hilbert matrices cannot be inverted with the standard precision of common numerical software!

4.15.3 Internal Methods for Matrices

A constructor that has been generated by means of Dom::Matrix or Dom::SquareMatrix contains many special functions for the corresponding data type. These are attached to the constructed objects as "methods". If M:=Dom::Matrix(ring) is a constructor and A:=M(..) is a matrix generated with this constructor, as described in Section 4.15.1, then the following methods are available:

- M::col(A,i) returns the ith column of A,
- M::delCol(A,i) removes the ith column from A,
- M::delRow(A,i) removes the ith row from A,
- M::dimen(A)[30] returns the dimension [m,n] of the $m \times n$ matrix A,
- M::random() returns a matrix with random entries,
- M::row(A,i) returns the ith row of A,
- M::swapCol(A,i,j) exchanges columns i and j,
- M::swapRow(A,i,j) exchanges rows i and j,
- M::tr(A) returns the trace $\sum_i A_{ii}$ of A,
- M::transpose(A) returns the transpose (A_{ji}) of $A = (A_{ij})$.

```
>> M := Dom::Matrix(): A := M([[x, 1], [2, y]]);
```

---

[30] The method dimen is renamed to matdim in MuPAD versions beyond 1.4.

```
+-      -+
|  x, 1  |
|        |
|  2, y  |
+-      -+
```

```
>> M::col(A, 1), M::delCol(A, 1), M::dimen(A);
```

```
+-   -+ +-    -+
|  x  | |  1  |
|     |, |     |, [2, 2]
|  2  | |  y  |
+-   -+ +-    -+
```

```
>> M::swapCol(A, 1, 2), M::tr(A), M::transpose(A);
```

```
+-      -+            +-      -+
|  1, x  |            |  x, 2  |
|        |, x + y,    |        |
|  y, 2  |            |  1, y  |
+-      -+            +-      -+
```

These methods, which are attached to the constructor M, are inherited by the generated object A:=M(..). In MuPAD versions up to 1.4, they may be called in the form A::method as well, for example:

```
>> A::col(A, 1), A::dimen(A), A::tr(A);
```

```
+-   -+
|  x  |
|     |, [2, 2], x + y
|  2  |
+-   -+
```

This is no longer possible in later versions, and you must access them as methods of the data type (i.e., the constructor). The function info returns a survey of these methods:

```
>> info(M);
```

```
-- Domain:
Dom::Matrix(Dom::ExpressionField(id, iszero))

-- Constructor:
Dom::Matrix

-- Super-Domains:
Dom::BaseDomain

-- Categories:
Cat::MatrixCat(Dom::ExpressionField(id, iszero)), Cat\
::SetCat

-- No Axioms.

-- Entries:
TeX, _concat, _divide, _index, _invert, _mult, _negat\
e, _plus, _power, _subtract, allAxioms, allCategories\
, allEntries, allSuperDomains, assign_elems, coeffRin\
g, coerce, col, concatMatrix, conjugate, convert, con\
vert_to, delCol, delRow, diff, dimen, equal, equiv, e\
xp, expand, expr, expr2text, float, gaussElim, getAxi\
oms, getCategories, getSuperDomain, has, hasProp, inf\
o, isDense, isSparse, iszero, length, map, mkDense, n\
ew, newThis, new_extelement, nonZeros, nops, norm, op\
, print, printElem, random, randomDimen, range_index,\
 row, setCol, setRow, set_index, stackMatrix, subs, s\
ubsex, subsop, swapCol, swapRow, testtype, tr, transp\
ose, undefinedEntries, whichEntry, zip
```

Below the subtitle **Entries**, you find a list of all methods of the domain. The call **?Dom::Matrix** returns a complete description of these methods.

4.15.4 The Libraries linalg and numeric

Besides the system function operating on matrices, the library[31] linalg contains a variety of other linear algebra functions:

---

[31] We refer to Chapter 3 for a description of general library organization, exporting, etc.

```
>> info(linalg);
```

```
    Library 'linalg': the linear algebra package
    Interface:
    linalg::addCol,          linalg::addRow,
    linalg::adjoint,         linalg::angle,
    linalg::basis,           linalg::charMatrix,
    linalg::charPolynomial,  linalg::cholesky,
      ...
    linalg::transpose,       linalg::vectorDimen,
    linalg::vectorPotential
```

Some of these functions, such as `linalg::col` or `linalg::delCol`, simply call the internal methods for matrices that we have described in Section 4.15.3, and hence do not add new functionality. However, `linalg` also contains many additional algorithms. The command `?linalg` yields a short description of all functions. You can find a detailed description of a function on the corresponding help page, for example:

```
>> ?linalg::det
```

```
    det -- determinant of a matrix

    Calling Sequence:

    det( A )

    Parameters:

    A -- square matrix

    Description:

      ...
```

You may use the full path name `library::function` to call a function:

```
>> A := Dom::Matrix()([[a, b], [c, d]]): linalg::det(A);
```

$$a\ d\ -\ b\ c$$

The characteristic polynomial $\det(x\,E - A)$ of this matrix is[32]

```
>> linalg::charPolynomial(A, x);
```

$$a\ d\ -\ b\ c\ +\ x^2\ +\ x\ (-\ a\ -\ d\ )$$

The eigenvalues are[33]

```
>> linalg::eigenValues(A);
```

```
{                                2     2 1/2
{ a    d   (4 b c - 2 a d + a  + d )
{ - + - - ----------------------------,
{ 2    2                   2


                            2     2 1/2 }
    a    d   (4 b c - 2 a d + a  + d )   }
    - + - + ---------------------------- }
    2    2                 2             }
```

The numerics library numeric (see ?numeric) contains many functions for numerical computations with matrices:

numeric::det	:	determinant computation
numeric::factorCholesky	:	Cholesky factorization
numeric::factorLU	:	*LU* factorization
numeric::factorQR	:	*QR* factorization
numeric::inverse	:	inversion
numeric::eigenvalues	:	eigenvalues
numeric::eigenvectors	:	eigenvalues and −vectors
numeric::singularvalues	:	singular values
numeric::singularvectors	:	singular values and vectors

---

[32] The function linalg::charPolynomial is renamed to linalg::charpoly in MuPAD versions beyond 1.4.

[33] The function linalg::eigenValues is renamed to linalg::eigenvalues in MuPAD versions beyond 1.4.

Partially, these routines work for matrices with symbolic entries of type Dom::ExpressionField and then are more efficient for large matrices than the linalg functions. However, the latter can handle arbitrary coefficient rings.

Exercise 4.39: Find the values of a, b, c for which the matrix $\begin{pmatrix} 1 & a & b \\ 1 & 1 & c \\ 1 & 1 & 1 \end{pmatrix}$ is not invertible.

Exercise 4.40: Consider the following matrices:

$$A = \begin{pmatrix} 1 & 3 & 0 \\ -1 & 2 & 7 \\ 0 & 8 & 1 \end{pmatrix}, \quad B = \begin{pmatrix} 7 & -1 \\ 2 & 3 \\ 0 & 1 \end{pmatrix}.$$

Let B^T be the transpose of B. Compute the inverse of $2\,A + B\,B^T$, both over the rational numbers and over the residue class ring modulo 7.

Exercise 4.41: Generate the $n \times n$ matrix

$$A_{ij} = \begin{cases} 0 & \text{for } i = j, \\ 1 & \text{for } i \neq j. \end{cases}$$

Compute its determinant, its characteristic polynomial, and its eigenvalues. For each eigenvalue, compute a basis of the corresponding eigenspace.

4.15.5 An Application

We want to compute the symbolic solution $a(t), b(t)$ of the system of second order differential equations

$$\frac{d^2}{dt^2}\,a(t) = 2\,c\,\frac{d}{dt}\,b(t) \;, \quad \frac{d^2}{dt^2}\,b(t) = -2\,c\,\frac{d}{dt}\,a(t) + 3\,c^2\,b(t) \;,$$

with an arbitrary constant c. If we let $a'(t) = \frac{d}{dt}a(t)$, $b'(t) = \frac{d}{dt}b(t)$, then the equations may be equivalently written as a system of first order differential equations in the variables $x(t) = (a(t), a'(t), b(t), b'(t))$:

$$\frac{d}{dt}\,x(t) = A\,x(t) \;,$$

where the matrix A is

$$A = \begin{pmatrix} 0 & 1 & 0 & 0 \\ 0 & 0 & 0 & 2c \\ 0 & 0 & 0 & 1 \\ 0 & -2c & 3c^2 & 0 \end{pmatrix} .$$

The solution of this system is given by applying the exponential matrix e^{tA} to the initial condition $x(0)$:

$$x(t) = e^{tA}x(0) .$$

Because of the symbolic constant c, we define the matrix A with coefficients in the domain Dom::ExpressionField of symbolic expressions, and require that all intermediate results be normalized automatically (Section 4.14):

```
>> M := Dom::Matrix(Dom::ExpressionField(normal, iszero));
>> A := M([[0,   1,    0,    0 ],
           [0,   0,    0,   2*c],
           [0,   0,    0,    1 ],
           [0, -2*c, 3*c^2, 0 ]]):
```

We use the function exp to compute $B = e^{tA}$:

```
>> B := exp(t*A);
```

```
    array(1..4, 1..4,
       (1, 1) = 1,
```

```
       . . .
```

$$(4, 4) = \frac{\exp(c\,(-t^2)^{1/2})}{2} + \frac{\exp(-c\,(-t^2)^{1/2})}{2}$$

```
    )
```

This result needs some simplification. We use the function rewrite (Section 9.1), which accepts an ordinary MuPAD expression as argument. But internally, the entries of the matrix B belong to a different coefficient ring:

```
>> domtype(B[1, 1]);
```

  Dom::ExpressionField(normal, iszero)

For that reason, we first convert all entries to ordinary MuPAD expressions via the constructor Dom::Matrix():

```
>> B := Dom::Matrix()(B):
```

Then we apply **rewrite** with the option **sincos** (Section 9.1) to replace the exponential functions by trigonometric functions:

```
>> B := map(B, rewrite, sincos):
```

Now the matrix B looks even more complicated than before, for example:

```
>> B[4, 4];
```

$$\frac{\cos(- I \, c \, (- t^2)^{1/2})}{2} + \frac{\cos(I \, c \, (- t^2)^{1/2})}{2} +$$

$$1/2 \, I \, \sin(- I \, c \, (- t^2)^{1/2}) +$$

$$1/2 \, I \, \sin(I \, c \, (- t^2)^{1/2})$$

We would like to simplify the subexpression (-t^2)^(1/2) further. Generally MuPAD assumes that an unknown is complex-valued, and does not simplify this expression due to the ambiguity of the complex square root. The command

```
>> assume(t >= 0):
```

tells MuPAD to treat t as a nonnegative real variable (Section 9.3). Then we find:

```
>> simplify((-t^2)^(1/2));

    I t
```

In the final step, we apply the composition of the functions `simplify` and `normal` to all entries of B. This generates the simplest form of this matrix:

```
>> B := map(B, normal@simplify);
```

```
+-                                                                        -+
|                                                                          |
|      - 3 c t + 4 sin(c t)                              - 2 cos(c t) + 2  |
|  1,  ---------------------,  6 c t - 6 sin(c t),       -----------------  |
|               c                        -.                     c          |
|                                                                          |
|  0,       4 cos(c t) - 3,    6 c - 6 c cos(c t),        2 sin(c t)       |
|                                                                          |
|           2 cos(c t) - 2                                 sin(c t)        |
|  0,       --------------,    - 3 cos(c t) + 4 ,         --------         |
|                 c                                           c            |
|                                                                          |
|  0,        -2 sin(c t),       3 c sin(c t),            cos(c t)          |
+-                                                                        -+
```

We assign a symbolic initial condition to the vector x(0):

```
>> x(0) := M([a(0), a'(0), b(0), b'(0)]):
```

Thus the desired symbolic solution of the system of differential equations is

```
>> x(t) := B*x(0):
```

The solution functions $a(t)$ and $b(t)$ with the symbolic initial conditions $a(0)$, $a'(0)$ (=D(a)(0)), $b(0)$, $b'(0)$ (=D(b)(0)) are:

```
>> a(t) := expand(expr(x(t)[1]));

    a(0) - 3 t D(a)(0) + 6 c t b(0) - 6 sin(c t) b(0) +

    2 D(b)(0)   2 cos(c t) D(b)(0)   4 sin(c t) D(a)(0)
    ---------- - ------------------ + ------------------
        c                c                   c
```

```
>> b(t) := expand(expr(x(t)[3]));
```

$$4 \ b(0) - 3 \ \cos(c \ t) \ b(0) - \frac{2 \ D(a)(0)}{c} +$$

$$\frac{2 \ \cos(c \ t) \ D(a)(0)}{c} + \frac{\sin(c \ t) \ D(b)(0)}{c}$$

Finally, we check that the above expressions really solve the system of differential equations:

```
>> expand(diff(a(t),t,t) - 2*c*diff(b(t),t)),
   expand(diff(b(t),t,t) + 2*c*diff(a(t),t) - 3*c^2*b(t));
```

```
0, 0
```

4.16 Polynomials

Computation with polynomials is an important task for a computer algebra system. Of course, you may realize a polynomial in MuPAD as an expression in the sense of Section 4.4 and use the standard arithmetic:

```
>> polynomialExpression := 1 + x + x^2:
>> expand(polynomialExpression^2);
```

$$2 \ x + 3 \ x^2 + 2 \ x^3 + x^4 + 1$$

However, there exists a special data type DOM_POLY, together with some kernel and library functions, which simplifies such computations and makes them more efficient.

4.16.1 Definition of Polynomials

The system function poly generates polynomials:

```
>> poly(1 + 2*x + 3*x^2);
```

```
      2
 poly(3 x  + 2 x + 1, [x])
```

Here we have supplied the expression $1 + 2x + 3x^2$ (of domain type DOM_EXPR) to poly, which converts this expression to a new object of domain type DOM_POLY. The indeterminate [x] is a fixed part of this data type. This is relevant for distinguishing between indeterminates and (symbolic) coefficients or parameters. For example, if you want to regard the expression $a_0 + a_1 x + a_2 x^2$ as a polynomial in x with coefficients a_0, a_1, a_2, then the above form of the call to poly does not yield the desired result:

```
>> poly(a0 + a1*x + a2*x^2);


      2
 poly(x  a2 + x a1 + a0, [x, a0, a1, a2])
```

This does not represent a polynomial in x but a "multivariate" polynomial in four indeterminates x, a_0, a_1, a_2. You can specify the indeterminates of a polynomial in form of a list as argument to poly. The system then considers all other symbolic identifiers as symbolic coefficients:

```
>> poly(a0 + a1*x + a2*x^2, [x]);


       2
 poly(a2 x + a1 x + a0, [x])
```

If you do not specify a list of indeterminates, then poly calls the function indets to determine all symbolic identifiers in the expression and interprets them as indeterminates of the polynomial:

```
>> indets(a0 + a1*x + a2*x^2, PolyExpr);

 {x, a0, a1, a2}
```

The distinction between indeterminates and coefficients is relevant for the representation of the polynomial:

```
>> expression := 1 + x + x^2 + a*x + PI*x^2 - b;
```

$$x - b + a\ x + x^2 + x^2\ PI + 1$$

```
>> poly(expression, [a, x]);
```

$$\text{poly}(a\ x + (PI + 1)\ x^2 + x + (1 - b),\ [a,\ x])$$

```
>> poly(expression, [x]);
```

$$\text{poly}((PI + 1)\ x^2 + (a + 1)\ x + (1 - b),\ [x])$$

You can see that MuPAD groups the coefficients of equal powers of the indeterminate together and sorts the terms according to falling exponents.

Instead of using an expression, you may also generate a polynomial by specifying a list of the nontrivial coefficients together with the respective exponents. The following command poly(list,[x]) generates the polynomial $\sum_{i=0}^{k} a_i\ x^{n_i}$ from the list

$$[[a_0, n_0], [a_1, n_1], \dots, [a_k, n_k]] :$$

```
>> list := [[1, 0], [a, 3], [b, 5]]: poly(list, [x]);
```

$$\text{poly}(b\ x^5 + a\ x^3 + 1,\ [x])$$

If you want to construct a multivariate polynomial in this way, specify lists of exponents for all variables:

```
>> poly([[3, [2, 1]], [2, [3, 4]]], [x, y]);
```

$$\text{poly}(2\ x^3\ y^4 + 3\ x^2\ y,\ [x,\ y])$$

Conversely, the function poly2list converts a polynomial to a list of coefficients and exponents:

```
>> poly2list(poly(b*x^5 + a*x^3 + 1, [x]) );
```

```
   [[b, 5], [a, 3], [1, 0]]
```

For more abstract computations, you may want to restrict the coefficients of a polynomial to a certain set (mathematically: a ring) which is represented by a special data structure in MuPAD. We have already seen typical examples of rings and their corresponding MuPAD domains in Section 4.14: the integers Dom::Integer, the rational numbers Dom::Rational, or the residue class ring Dom::IntegerMod(n) of integers modulo n. You may specify the coefficient ring as argument to poly:

```
>> poly(x+1, [x], Dom::Integer);
```

```
   poly(x + 1, [x], Dom::Integer)
```

```
>> poly(2*x-1/2, [x], Dom::Rational);
```

```
   poly(2 x - 1/2, [x], Dom::Rational)
```

```
>> poly(4*x + 11, [x], Dom::IntegerMod(3));
```

```
   poly(x + 2, [x], Dom::IntegerMod(3))
```

Note that in the last example, the system has automatically simplified the coefficients according to the rules for computing with integers[34] modulo 3:

```
>> 4 mod 3, 11 mod 3;
```

```
   1, 2
```

In the following example, poly converts the coefficients to floating-point numbers, as specified by the third argument:

---

[34] You may also use this coefficient ring in the form poly(4*x+11,[x],IntMod(3)). Then the integers modulo 3 are represented by $-1, 0, 1$ and not, as for Dom::IntegerMod(3), by $0, 1, 2$. Polynomial arithmetic is much faster, when you use IntMod(3).

```
>> poly(PI*x - 1/2, [x], Dom::Float);
```

```
   poly(3.141592653 x - 0.5, [x], Dom::Float)
```

If no coefficient ring is specified, then MuPAD by default uses the ring **Expr** which symbolizes arbitrary MuPAD expressions. In this case, you may use symbolic identifiers as coefficients:

```
>> polynomial := poly(a + x + b*y, [x, y]);
   op(polynomial);
```

```
   poly(x + b y + a, [x, y])
```

```
   a + x + b y, [x, y], Expr
```

We summarize that a MuPAD polynomial comprises three parts:

1. a polynomial expression of the form $\sum a_{i_1 i_2..} x_1^{i_1} x_2^{i_2} \cdots$,
2. a list of indeterminates $[x_1, x_2, \ldots]$,
3. the coefficient ring.

These are the three operands of a MuPAD polynomial p, which can be accessed via op(p,1), op(p,2), and op(p,3), respectively. Thus you may convert a polynomial to a mathematically equivalent expression of domain type DOM_EXPR by

```
>> expression := op(polynomial, 1):
```

However, you should preferably use the system function **expr**, which can convert various domain types such as polynomials to expressions:

```
>> polynomial := poly(x^3 + 5*x + 3);
```

```
          3
   poly(x  + 5 x + 3, [x])
```

```
>> op(polynomial, 1) = expr(polynomial);
```

```
        3                3
   5 x + x  + 3 = 5 x + x  + 3
```

4.16.2 Computing with Polynomials

The function **degree** determines the degree of a polynomial:

```
>> p := poly(1 + x + a*x^2*y, [x, y]):
>> degree(p, x), degree(p, y);
```

   2, 1

If you do not specify the name of an indeterminate as second argument, then **degree** returns the "total degree":

```
>> degree(p), degree(poly(x^27 + x + 1));
```

   3, 27

The function **coeff** extracts coefficients from a polynomial:

```
>> p := poly(1 + a*x + 7*x^7, [x]):
>> coeff(p, 1), coeff(p, 2), coeff(p, 8);
```

   a, 0, 0

For multivariate polynomials, the coefficient of a power of one particular indeterminate is again a polynomial in the remaining indeterminates:

```
>> p := poly(1 + x + a*x^2*y, [x, y]):
>> coeff(p, y, 0), coeff(p, y, 1);
```

$$
poly(x + 1, [x]), \quad poly(a\ x^2, [x])
$$

The standard operators +, -, * and ^ work for polynomial arithmetic as well:

```
>> p := poly(1 + a*x^2, [x]): q := poly(b + c*x, [x]):
>> p + q, p - q, p*q, p^2;
```

$$poly(a\ x^2 + c\ x + (b + 1),\ [x]),$$

$$poly(a\ x^2 + (-c)\ x + (1 - b),\ [x]),$$

$$poly((a\ c)\ x^3 + (a\ b)\ x^2 + c\ x + b,\ [x]),$$

$$poly(a^2\ x^4 + (2\ a)\ x^2 + 1,\ [x])$$

The function `divide` performs a "division with remainder":

```
>> p := poly(x^3 + 1): q := poly(x^2 - 1): divide(p, q);

    poly(x, [x]), poly(x + 1, [x])
```

The result is a sequence with two operands: the quotient and the remainder of the division. If we set

```
>> quotient := op(divide(p,q),1):
   remainder := op(divide(p,q),2):
```

then p = quotient*q + remainder:

```
>> quotient*q + remainder;

    poly(x^3 + 1, [x])
```

The polynomial denoted by `remainder` is of lower degree than q, which makes the decomposition p = `quotient*q` + `remainder` unique. Dividing two polynomials by means of the usual division operator / is only allowed in the special case when the remainder that `divide` would return vanishes:

```
>> p := poly(x^2 - 1): q := poly(x - 1): p/q;

    poly(x + 1, [x])
```

```
>> p := poly(x^2 + 1): q := poly(x - 1): p/q;
```

```
FAIL
```

Note that the arithmetic operators process only polynomials of exactly identical types:

```
>> poly(x + y, [x, y]) + poly(x^2, [x, y]),
   poly(x) + poly(x, [x], Expr);
```

```
        2
  poly(x  + x + y, [x, y]), poly(2 x, [x])
```

Both the list of indeterminates and the coefficient ring must agree, otherwise the system returns the input as a symbolic expression:

```
>> poly(x + y, [x, y]) + poly(x^2, [x]);
```

```
        2
  poly(x , [x]) + poly(x + y, [x, y])
```

```
>> poly(x) + poly(x, Dom::Integer);
```

```
  poly(x, [x]) + poly(x, [x], Dom::Integer)
```

The polynomial arithmetic performs coefficient additions and multiplications according to the rules of the coefficient ring:

```
>> p := poly(4*x + 11, [x], Dom::IntegerMod(3)):
>> p; p + p; p*p;
```

```
  poly(x + 2, [x], Dom::IntegerMod(3)),

  poly(2 x + 1, [x], Dom::IntegerMod(3)),

        2
  poly(x + x + 1, [x], Dom::IntegerMod(3))
```

The standard operator * does not work immediately for multiplying a polynomial by a scalar; you need to convert the scalar factor to a polynomial first:

```
>> p := poly(x^2 + y):
>> scalar*p; poly(scalar, op(p, 2..3))*p;
```

$$poly(x^2 + y, [x, y]) \ scalar$$

$$poly(scalar \ x^2 + scalar \ y, [x, y])$$

Here we have ensured that the polynomial generated from the scalar factor is of the same type as p by passing op(p,2..3) (=[x,y],Expr) as further arguments to poly. Alternatively, the function multcoeffs multiplies all coefficients of a polynomial by a scalar factor:

```
>> multcoeffs(p, scalar);
```

$$poly(scalar \ x^2 + scalar \ y, [x, y])$$

The function mapcoeffs applies an arbitrary function to all coefficients of a polynomial:

```
>> p := poly(2*x^2 + 3*y): mapcoeffs(p, f);
```

$$poly(f(2) \ x^2 + f(3) \ y, [x, y])$$

This yields another way of multiplication by a scalar:

```
>> mapcoeffs(p, _mult, scalar);
```

$$poly((2 \ scalar) \ x^2 + (3 \ scalar) \ y, [x, y])$$

Another important operation is the evaluation of a polynomial at a point (computing the image value). The function evalp achieves this:

```
>> p := poly(x^2 + 1, [x]):
   evalp(p, x = 2), evalp(p, x = x + y);
```

$$5, \ (x + y)^2 + 1$$

This computation is also valid for multivariate polynomials and yields a polynomial in the remaining indeterminates or, for a univariate polynomial, an element of the coefficient ring:

```
>> p := poly(x^2 + y):
>> q := evalp(p, x = 0); evalp(q, y = 2);
```

```
   poly(y, [y])
```

```
   2
```

Equivalently, you may also regard a polynomial as a function of the indeterminates and call this function with arguments:

```
>> p(2, z);
```

```
   z + 4
```

A variety of MuPAD functions accepts polynomials as input. An important operation is factorization, which is performed according to the rules of the coefficient ring. The functions `factor` and `Factor`[35] return the factors in different output formats:

```
>> factor(poly(x^3 - 1));
```

$$[1, \ poly(x - 1, \ [x]), \ 1, \ poly(x^2 + x + 1, \ [x]), \ 1]$$

```
>> Factor(poly(x^2 + 1, Dom::IntegerMod(2)));
```

$$poly(x + 1, \ [x], \ Dom::IntegerMod(2))^2$$

---

[35] In MuPAD versions beyond 1.4, the function `Factor` is obsolete, and the function `factor` returns a MuPAD object of domain type `Factored`. It is printed on the screen in the same format as the output of `Factor`. Internally, however, the irreducible factors and the exponents are still stored in form of a list, and you can use them via `op` or indexed accesses. See the help pages `?factor` and `?Factored` for details.

The function D differentiates polynomials:

```
>> D(poly(x^7 + x + 1));
```

$$poly(7\ x^6 + 1,\ [x])$$

Equivalently, you may also use diff(polynomial,x). Integration also works for polynomials[36]:

```
>> p := poly(x^7 + x + 1): int(p, x);
```

$$poly(1/8\ x^8 + 1/2\ x^2 + x,\ [x])$$

The function gcd computes the greatest common divisor of polynomials:

```
>> p := poly((x + 1)^2*(x + 2)):
>> q := poly((x + 1)*(x + 2)^2):
>> Factor(gcd(p, q));
```

```
poly(x + 2, [x]) poly(x + 1, [x])
```

The internal representation of a polynomial stores only those powers of the indeterminates with non-vanishing coefficients. This is particularly advantageous for "sparse" polynomials of high degree and makes arithmetic with such polynomials efficient. The function nterms returns the number of nontrivial terms of a polynomial. The function nthmonomial extracts individual monomials (coefficient times powers of the indeterminates), nthcoeff and nthterm return the appropriate coefficient and product of powers of the indeterminates, respectively:

```
>> p := poly(a*x^100 + b*x^10 + c, [x]):
>> nterms(p), nthmonomial(p, 2),
   nthcoeff(p, 2), nthterm(p, 2);
```

$$3,\ poly(b\ x^{10},\ [x]),\ b,\ poly(x^{10},\ [x])$$

---

[36] Before MuPAD version 1.4.2 you have to convert a polynomial to an expression via expr to integrate it.

Table 4.3 is a summary of the operations for polynomials discussed above. Section "Functions for Polynomials" of the MuPAD Quick Ref-

`+, -, *, ^`	:	arithmetic
`coeff`	:	extract coefficients
`degree`	:	polynomial degree
`diff, D`	:	differentiation
`divide`	:	division with remainder
`evalp`	:	evaluation
`expr`	:	conversion to expression
`Factor, factor`	:	factorization
`gcd`	:	greatest common divisor
`mapcoeffs`	:	apply a function
`multcoeffs`	:	multiplication by a scalar
`nterms`	:	number of nonzero coefficients
`nthcoeff`	:	nth coefficient
`nthmonomial`	:	nth monomial
`nthterm`	:	nth term
`poly`	:	construct a polynomial
`poly2list`	:	conversion to list

Table 4.3. MuPAD functions operating on polynomials

erence [Oev 98] lists further functions for polynomials in the standard library. The `groebner` library comprises functions for handling multi-variate polynomial ideals (`?groebner`).

Exercise 4.42: Consider the polynomials $p = x^7 - x^4 + x^3 - 1$ and $q = x^3 - 1$. Compute $p - q^2$. Does q divide p? Factor p and q.

Exercise 4.43: A polynomial is called irreducible (over a coefficient field), if it cannot be factored into a product of more than one nonconstant polynomials. The function `irreducible` tests a polynomial for irreducibility. Find all irreducible quadratic polynomials $a\,x^2 + b\,x + c$, $a \neq 0$ over the field of integers modulo 3.

4.17 Null Objects: `null()`, `NIL`, `FAIL`, and `undefined`

There are several objects representing the "void" in MuPAD. First, we have the "empty sequence" generated by `null()`. It is of domain type

DOM_NULL and generates no output on the screen. System functions such as `reset` (Section 14.3) or `print` (Section 13.1.1), which cannot return mathematically useful values, return this MuPAD object instead:

```
>> a := reset(): b := print("hello"):
```

```
                    "hello"
```

```
>> domtype(a), domtype(b);
```

```
   DOM_NULL, DOM_NULL
```

The object `null()` is particularly useful in connection with sequences (Section 4.5). The system automatically removes this object from sequences, and you can use it, for example, to remove sequence entries selectively:

```
>> Seq := a, b, c: Seq := eval(subs(Seq, b = null()));
```

```
   a, c
```

Here we have used the substitution command `subs` (Chapter 6) to replace b by `null()`.

The MuPAD object NIL, which is distinct from `null()`, intuitively means "no value". Some system functions return the NIL object when you call them with arguments for which they need not compute anything. A typical example is the function `_if`, which is usually called in form of an `if` statement (Chapter 17):

```
>> condition := FALSE: if condition then x := 1 end_if;
```

```
   NIL
```

The MuPAD object FAIL intuitively means "I could not find a value". System functions return this object when there is no meaningful result for the given input parameters. In the following example, we try to compute the inverse of a singular matrix:

```
>> A := Dom::Matrix()([[1, 2], [2, 4]]);
```

```
+-       -+
|  1, 2  |
|        |
|  2, 4  |
+-       -+
```

```
>> A^(-1);
```

```
FAIL
```

Another object with a similar meaning is **undefined**. For example, the MuPAD function **limit** returns this object when the requested limit does not exist:

```
>> limit(1/x, x = 0);
```

```
undefined
```

5. Evaluation and Simplification

5.1 Identifiers and Their Values

Consider:

```
>> unassign(x, a): y := a + x;

   a + x
```

Since the identifiers **a** and **x** only represent themselves, the "value" of **y** is the symbolic expression **a+x**. We have to distinguish carefully between the identifier **y** and its value. More precisely, the *value* of an identifier denotes the MuPAD object that the system computes by evaluation and simplification of the right hand side of the assignment `identifier :=value` *at the time of assignment.*

Note that in the above example the value of **y** is composed of the symbolic identifiers **a** and **x**, which may be assigned values at a later time. For example, if we assign the value 1 to the identifier **a**, then **a** is replaced by its value 1 in the expression **a+x**, and the call **y** returns **x+1**:

```
>> a := 1: y;

   x + 1
```

We say that the *evaluation* of the identifier **y** returns the result **x+1**, but its *value* is still **a+x**:

> We distinguish between an identifier, its value, and its evaluation: the *value* denotes the evaluation *at the time of assignment*, a later *evaluation* may return a different *"current value"*.

If we now assign the value 2 to x, then both a and x are replaced by their values at the next evaluation of y. Hence we obtain the sum 2+1 as a result, which MuPAD automatically simplifies to 3:

```
>> x := 2: y;
```

```
                     3
```

The *evaluation* of y now returns the number 3; its *value* is still $a + x$.

It is reasonable to say that the value of y is the result at the time of assignment. Namely, if we delete the values of the identifiers a and x in the above example, then the evaluation of y yields its original value immediately after the assignment:

```
>> unassign(a, x): y;
```

```
                   a + x
```

If a or x already have a value *before* we assign the expression a+x to y, then the following happens:

```
>> x := 1: y := a + x: y;
```

```
                   a + 1
```

At the time of assignment, y is assigned the evaluation of a+x, i.e., a+1. Indeed, this is now the *value* of y, which contains no reference to x:

```
>> unassign(x): y;
```

```
                   a + 1
```

Here are some further examples for this mechanism. We first assign the rational number 1/3 to x, then we assign the object [x, x^2, x^3] to the identifier list. In the assignment the system evaluates the right hand side and automatically replaces the identifier x by its value. Thus the identifier list gets the value [1/3, 1/9, 1/27], and not [x, x^2, x^3], at the time of assignment:

```
>> x := 1/3: list := [x, x^2, x^3];
```

$$[1/3,\ 1/9,\ 1/27]$$

```
>> unassign(x): list;
```

$$[1/3,\ 1/9,\ 1/27]$$

MuPAD applies the same evaluation scheme to symbolic function calls:

```
>> unassign(f): y := f(PI);
```

$$f(PI)$$

After the assignment

```
>> f := sin:
```

we obtain the evaluation

```
>> y;
```

$$0$$

When evaluating y the system replaced the identifier f by its value, which is the value of the identifier sin. This is a procedure which is executed when y is evaluated and returns sin(PI) as 0.

5.2 Complete, Incomplete, and Enforced Evaluation

We consider once again the first example from the previous section. There we have assigned the expression a+x to the identifier y, and a and x did not have a value:

```
>> unassign(a, x): y := a + x: a := 1: y;
```

$$x + 1$$

We now explain in greater detail how MuPAD performs the final evaluation.

First ("level 1") the evaluator considers the value a+x of y. Since this value contains identifiers x and a, a second evaluation step ("level 2") is necessary to determine the value of these identifiers. The system recognizes that a has the value 1, while x has no value (and thus mathematically represents an unknown). Now the system's arithmetic combines these results to x+1, and this is the evaluation of y. Figures 5.1–5.3 illustrate this process. A box represents an identifier and its value (or ·, respectively, if it has no value). An arrow represents one evaluation step.

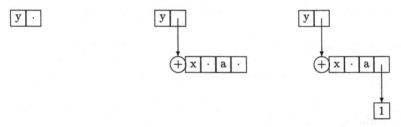

Figure 5.1. The identifier y without a value.

Figure 5.2. After the assignment y:=a+x.

Figure 5.3. After the assignment a:=1, we finally obtain x+1 as the evaluation of y.

In analogy to the expression trees for representing symbolic expressions (Section 4.4.2), you may imagine the process of evaluation as an *evaluation tree*, whose vertices are expressions with symbolic identifiers, with branches pointing to the corresponding values of these identifiers. The system traverses this tree until there are either no identifiers left or all remaining identifiers have no value.

The user may control the levels of this tree via the system function level. We look at an example:

```
>> unassign(a, b, c): x := a + b: a := b + 1: b := c:
```

The evaluation tree for x is:

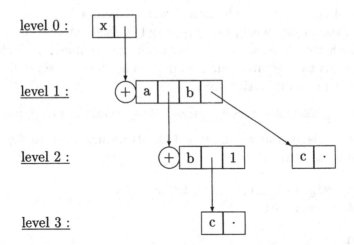

The identifier x forms the top level (the root, level 0) of its own evaluation tree:

```
>> level(x, 0);

    x
```

The next level 1 determines the value of x:

```
>> level(x, 1);

    a + b
```

At the following level 2, a and b are replaced by their values b+1 and c, respectively:

```
>> level(x, 2);

    b + c + 1
```

The remaining b is replaced by its value c only in the next level 3:

```
>> level(x, 3);

    2 c + 1
```

We call the type of evaluation described here a *complete evaluation*. This means that identifiers are replaced by their values recursively until no further evaluations are possible. The environment variable LEVEL, which has the default value 100, determines how far MuPAD descends at most in an evaluation tree.

In interactive mode, MuPAD always evaluates completely!

More precisely, this means that MuPAD evaluates up to depth LEVEL in interactive mode[1].

```
>> unassign(a0, a1, a2): LEVEL := 2:
>> a0 := a1: a0;

    a1

>> a1 := a2: a0;

    a2
```

Up to now, the evaluation tree for a0 has depth 2, and the LEVEL value of 2 achieves a complete evaluation. However, in the next step the value of a2 is not taken into account:

```
>> a2 := a3: a0;

    a2

>> unassign(LEVEL):
```

As soon as MuPAD realizes that the current evaluation level exceeds the value of the environment variable MAXLEVEL (whose default value is 100), then it supposes to be in an infinite loop and aborts the evaluation with an error message:

```
>> MAXLEVEL := 2: a0;

    Error: Recursive definition [See ?MAXLEVEL]
```

---

[1] You must not confuse this with the effect of a system function call, which may return a *not completely evaluated object*, such as subs (Chapter 6). The call subs(sin(x),x=0), for example, returns sin(0) and not 0! The functionality of subs is to perform a substitution and to return the resulting object without further evaluation.

```
>> unassign(MAXLEVEL):
```

We now present some important exceptions to the rule of complete evaluation. The calls `last(i)`, `%i`, or `%` (Chapter 12) do not lead to an evaluation! We consider the example

```
>> unassign(x): [sin(x), cos(x)]: x := 0:
```

Now `%2` accesses the list without evaluating it:

```
>> %2;
```

```
   [sin(x), cos(x)]
```

However, you can enforce evaluation by means of `eval`

```
>> eval(%);
```

```
   [0, 1]
```

Compare this to the following statements, where requesting the identifier `list` causes the usual complete evaluation:

```
>> unassign(x): list := [sin(x), cos(x)]: x := 0: list;
```

```
   [0, 1]
```

Arrays of domain type `DOM_ARRAY` are always evaluated with level 1:

```
>> unassign(a, b): A := array(1..2, [a, b]):
>> b := a: a := 1: A;
```

```
   +-     -+
   | a, b |
   +-     -+
```

As you can see, the call of `A` returns the value (the array), but does not replace `a,b` by their values. You can enforce the evaluation of the entries by[2]:

---

[2] In MuPAD versions beyond 1.4, these commands have no effect. However, you can enforce the complete evaluation of all array entries via `map(A, eval)`.

```
>> map(A, level, 1), map(A, level, 2);
```

```
+-      -+ +-      -+
| 1, a |, | 1, 1 |
+-      -+ +-      -+
```

Note that in contrast to the above behavior, the indexed access of an individual entry is evaluated completely:

```
>> A[1], A[2];
```

```
    1, 1
```

Matrices (of domain type Dom::Matrix(..)), tables (of domain type DOM_TABLE), and polynomials (DOM_POLY) are treated in the same way. Moreover, within procedures MuPAD always evaluates only up to level 1 (Section 18.10). If this is not sufficient, then you may control this behavior explicitly by means of level.

The command hold(object) corresponds to level(object,0) and prevents the evaluation of object. This may be desirable in many situations. For example, the following attempt to generate the sequence X[1],..,X[8] does not work in MuPAD version 1.4:

```
>> i := 2: X[i] $ i = 1..8;
```

```
    Error: Illegal argument [_seqgen]
```

The reason is that the right hand side i=1..8 of the $ operator is evaluated to 2=1..8, and the symbolic loop variable i is lost. The function hold protects the loop variable from evaluation. This avoids the problem above[3]:

```
>> X[i] $ hold(i) = 1..8;
```

```
    X[1], X[2], X[3], X[4], X[5], X[6], X[7], X[8]
```

MuPAD does not evaluate the left hand side of the $ operator, and the loop variable need not be encapsulated there. The following function,

---

[3] In later MuPAD versions, the $ operator automatically protects the loop variable. Then the use of hold is not only unnecessary, it is invalid!

which cannot be executed for symbolical arguments, yields another example where the (premature) evaluation is undesirable:

```
>> absValue := X -> (if X >= 0 then X else -X end_if):
>> absValue(X);
```

```
   Error: Can't evaluate to boolean [_less]
```

If you want to plot this function by using the function plotfunc[4] (Chapter 11), then specifying the symbolic expression absValue(X) in the form:

```
>> plotfunc(absValue(X), X = -1..1);
```

provokes an error message. If you delay the evaluation of absValue(X) by:

```
>> plotfunc(hold(absValue)(X), X = -1..1);
```

then this produces the desired figure. The reason is that plotfunc internally substitutes numerical values for X, for which absValue can be evaluated without problems.

Here is another example: like most MuPAD functions, the function domtype first evaluates its argument, so that the command domtype(object) returns the domain type of the *evaluation* of object:

```
>> x := 1: y := 1: x, x + y, sin(0), sin(0.1);
```

```
   1, 2, 0, 0.09983341664
```

```
>> domtype(x), domtype(x + y), domtype(sin(0)),
   domtype(sin(0.1));
```

```
   DOM_INT, DOM_INT, DOM_INT, DOM_FLOAT
```

Using hold, you obtain the domain types of the objects themselves: x is an identifier, x+y is an expression, and sin(0) and sin(0.1) are function calls and hence expressions as well:

---

[4] The function plotfunc is replaced by plotfunc2d and plotfunc3d in MuPAD versions beyond 1.4.

```
>> domtype(hold(x)), domtype(hold(x + y)),
   domtype(hold(sin(0))), domtype(hold(sin(0.1)));

   DOM_IDENT, DOM_EXPR, DOM_EXPR, DOM_EXPR
```

The commands ?level and ?hold provide further information from the corresponding help pages.

Exercise 5.1: What are the *values* of the identifiers x, y, and z after the following statements? What is the *evaluation* of the last statement in each case?

```
>> x := a1: a1 := b1: a1 := c1: x;
>> a2 := b2: y := a2: b2 := c2: y;
>> b3 := a3: z := b3: a3 := 10: z;
```

Predict the results of the following statement sequences:

```
>> u1 := v1: v1 := w1: w1 := u1: u1;
>> u2 := v2: u2 := u2^2 - 1: u2;
```

5.3 Automatic Simplification

MuPAD automatically simplifies many objects, such as certain function calls or arithmetic expressions with numbers:

```
>> sin(15*PI), exp(0), (1 + I)*(1 - I);

   0, 1, 2
```

The same holds true for arithmetic expressions containing the object infinity:

```
>> 2*infinity - 5;

   infinity
```

Such simplifications are performed by MuPAD's *internal simplifier*, which is a part of the system kernel. It very efficiently reduces the complexity of expressions:

```
>> cos(1 + exp((-1)^(1/2)*PI));
```

$$1$$

The user can neither control nor extend the internal simplifier.

In most cases, MuPAD does *not* automatically simplify expressions. The reason is that the system generally cannot decide which is the most reasonable way of simplification. For example, consider the following expression which is not simplified:

```
>> y := (-4*x + x^2 + x^3 - 4)*(7*x - 5*x^2 + x^3 - 3);
```

$$(- 4 x + x^2 + x^3 - 4) (7 x - 5 x^2 + x^3 - 3)$$

Naturally, you can expand this expression, which may be sensible, for example, before computing its symbolic integral:

```
>> expand(y);
```

$$- 16 x - 11 x^2 + 20 x^3 - 2 x^4 - 4 x^5 + x^6 + 12$$

However, if you are interested in the roots of the polynomial, then you had better compute its linear factors:[5]

```
>> Factor(y);
```

$$(x - 2) (x + 2) (x + 1) (x - 3) (x - 1)^2$$

No universal answer is possible to the question of which of the two representations is "simpler". Depending on the intended application, you can selectively apply appropriate system functions (such as expand or Factor) to simplify an expression.

There is another argument against automatic simplification. The symbol f, for example, might represent a bounded function, for which the limit $\lim_{x \to 0} x f(x)$ is 0. However, simplifying this expression to 0

---

[5] The function Factor is replaced by factor in MuPAD versions beyond 1.4; see the footnote on page 135.

can be wrong for functions f with a singularity at the origin (such as $f(x) = 1/x$)! Thus automatic simplifications such as $0 \cdot f(0) = 0$ are questionable as long as the system has no additional knowledge about the symbols involved. In general, MuPAD cannot know which rule can be applied and which must not. Now you might object that MuPAD should perform no simplifications at all instead of wrong ones. Unfortunately this is not reasonable either, since in symbolic computations, expressions tend to grow very quickly, and this seriously affects the performance of the system. In fact, MuPAD always simplifies an expression of the form 0*y to 0, except when the value of y is infinity. You should always keep in mind that such a simplified result may be wrong in extreme cases.

Another example is the solution of the equation $x/x = 1$ for $x \neq 0$. In this case, MuPAD 1.4.2 returns:

```
>> solve(x/x = 1);
```

```
    {0 = 0}
```

MuPAD's equation solver solve (Chapter 8) returns simplified ("solved") equations that are equivalent to the original ones. In the above example, MuPAD claims that the original equation is equivalent to the trivial equation $0 = 0$. This would imply that *arbitrary* values of x yield a solution of the equation $x/x = 1$. The system first automatically simplifies the expression x/x to 1, and then in fact solves the equation $1 = 1$ by—reasonably—simplifying it to $0 = 0$. The exceptional case $x = 0$, for which the original problem makes no sense, is completely ignored in the simplified output!

For that reason, MuPAD performs only some simplifications automatically, and you must explicitly request other simplifications yourself. For this purpose MuPAD provides a variety of functions, some of which are described in Section 9.2.

6. Substitution: subs, subsex, and subsop

All MuPAD objects consist of operands (Section 4.1). An important feature of a computer algebra system is that it can replace these building blocks by new values. For that purpose, MuPAD provides the functions subs, subsex (short for: substitute expression), and subsop (short for: substitute operand).

The command subs(object,Old=New) replaces all occurrences of the subexpression Old in object by the value New:

```
>> f := a + b*c^b: g := subs(f, b = 2):  f, g;
```

$$a + b\,c^{b} \;,\; a + 2\,c^{2}$$

You see that subs returns the result of the substitution,, but the identifier f remains unchanged. If you represent a map F by the expression $f = F(x)$, then you may use subs to evaluate the function at some point:

```
>> f := 1 + x + x^2:
>> subs(f, x = 0), subs(f, x = 1),
   subs(f, x = 2), subs(f, x = 3);
```

```
1, 3, 7, 13
```

The output of the subs command is subjected to the usual simplifications of MuPAD's internal simplifier. In the above example, the call subs(f,x=0) produces the object 1+0+0^2, which is automatically simplified to 3. You must not confuse this with *evaluation* (Chapter 5), where in addition all identifiers in an expression are replaced by their values.

The function subs performs a substitution. The system only simplifies the resulting object, but does not evaluate it upon return!

In the following example

```
>> f := x + sin(x): g := subs(f, x = 0);
```

```
     sin(0)
```

the identifier sin for the sine function is not replaced by the corresponding MuPAD function, which would return sin(0)=0. Only the next call to g performs a complete evaluation:

```
>> g;
```

```
     0
```

You can enforce evaluation by using eval:

```
>> eval(subs(f, x = 0));
```

```
     0
```

You may replace arbitrary MuPAD objects by substitution. In particular, you can substitute functions or procedures as new values:

```
>> eval(subs(h(a + b), h = (x -> (1 + x^2))));
```

```
            2
     (a + b)  + 1
```

If you want to replace a system function, enclose it in a hold command:

```
>> f := subs(sin(a + b), hold(sin) = x -> (x - x^3/3)): f;
```

```
                3
         (a + b)
     a + b - ---------
                3
```

You can also replace more complex subexpressions:

```
>> subs(sin(x)/(sin(x) + cos(x)), sin(x) + cos(x) = 1);
```

```
    sin(x)
```

However, you should be careful with such substitutions: the command
subs(object,Old=New) replaces all those occurrences of the expression Old *that can be found by means of op*. This explains why nothing happens in the following example:

```
>> subs(a + b + c, a + b = 1), subs(a*b*c, a*b = 1);
```

```
    a + b + c,  a b c
```

Here the sum a+b and the subproduct a*b are *not* operands of the corresponding expressions. In contrast, we find:

```
>> f := a + b + sin(a + b): subs(f, a + b = 1);
```

```
    a + b + sin(1)
```

Again, you cannot obtain the subexpression a+b by means of op. However, the argument of the sine is the suboperand op(f, [3,1]) (see Sections 4.1 and 4.4.3), and hence it is replaced by 1. In contrast to subs, the function subsex also replaces subexpressions in sums and products:

```
>> subsex(f, a + b = x + y), subsex(a*b*c, a*b = x + y);
```

```
    x + y + sin(x + y),  c (x + y)
```

This kind of substitution requires a closer analysis of the expression tree, and hence subsex is much slower than subs for large objects. When replacing more complex subexpressions, you should not be mislead by the screen output of expressions:

```
>> f := a/(b*c);
```

```
   a
  ---
  b c
```

```
>> subs(f, b*c = New), subsex(f, b*c = New);
```

```
   a    a
  ---, ---
  b c  b c
```

If you look at the operands of f, then you see that the expression tree does not contain the product b*c, which explains why no substitution took place:

```
>> op(f);
```

```
      1  1
   a, -, -
      b  c
```

You can perform several substitutions with a single call of subs:

```
>> subs(a + b + c, a = A, b = B, c = C);
```

```
   A + B + C
```

This is equivalent to the nested call

```
>> subs(subs(subs(a + b + c, a = A), b = B), c = C):
```

Thus we obtain:

```
>> subs(a + b^2, a = b, b = a);
```

```
        2
   a + a
```

First MuPAD replaces a by b, yielding b+b^2. Then it substitutes a for b in this new expression and returns the above result. In contrast, you may achieve a *simultaneous substitution* by specifying the substitution equations in form of a list or a set:

```
>> subs(a + b^2, [a = b, b = a]),
   subs(a + b^2, {a = b, b = a});
```

$$b + a^2, \quad b + a^2$$

The output of the equation solver **solve** (Chapter 8) supports the functionality of **subs**. In general, **solve** returns sets of equations, which may be fed in **subs** immediately:

```
>> equations := {x + y = 2, x - y = 1}:
>> solution := solve(equations, {x, y});
```

$$\{\{y = 1/2, x = 3/2\}\}$$

```
>> subs(equations, op(solution, 1));
```

$$\{1 = 1, 2 = 2\}$$

The function **subsop** provides another variant of substitution: **subsop(object,i=New)** selectively replaces the ith operand of the object by the value **New**:

```
>> subsop(2*c + a^2, 2 = d^5);
```

$$2 c + d^5$$

Here, we have replaced the second operand a^2 of the sum by d^5. In the following example, we first replace the exponent of the second term (this is the operand [2,2] of the sum), and then the first term:

```
>> subsop(2*c + a^2, [2, 2] = 4, 1 = x*y);
```

$$x y + a^4$$

In the following expression, we first replace the first term, yielding the expression x*y+c^2. Then we substitute z for the second factor of the first term (which now is y):

```
>> subsop(a*b + c^2, 1 = x*y, [1, 2] = z);
```

$$x z + c^2$$

The expression a+2 is a symbolic sum, which has a 0th operand. Namely, this is the system function _plus, for generating sums:

```
>> op(a + 2, 0);
```

```
    _plus
```

You can replace this operand by any other function (for example, by the system function _mult which multiplies its arguments):

```
>> subsop(a + 2, 0 = _mult);
```

```
    2 a
```

When using subsop, you need to know the position of the operand that you want to replace. Nonetheless, you should be cautious, since the system may change the order of the operands when this is mathematically valid (for example, in sums, products, or sets):

```
>> set := {a, b, c^2, sin(1 + a)};
```

$$\{c^2, a, b, \sin(a + 1)\}$$

If you use subs, then you need not know the position of the subexpression. Another difference between subs and subsop is that subs traverses the expression tree of the object *recursively*, and thus also replaces suboperands:

```
>> subs(set, a = a^2);
```

$$\{c^2, b, \sin(a^2 + 1), a^2\}$$

Exercise 6.1: Does the command subsop(b+a,1=c) replace the identifier b by c?

Exercise 6.2: The commands

```
>> unassign(f): g := diff(f(x)/diff(f(x),x), x $ 5);
```

```
  25 diff(f(x), x, x) diff(f(x), x, x, x, x)
  --------------------------------------------- -
                          2
              diff(f(x), x)


   4 diff(f(x), x, x, x, x, x)
  ---------------------------- -  ...
          diff(f(x), x)
```

generate a lengthy expression containing symbolic derivatives. Make this expression more readable by replacing these derivatives by simpler names $f_0 = f(x)$, $f_1 = f'(x)$, etc.

7. Differentiation and Integration

We have already used MuPAD's commands for differentiation and integration. Since they are important, we recapitulate the usage of these routines here.

7.1 Differentiation

The call diff(expression,x) computes the derivative of the expression with respect to the unknown x:

```
>> diff(sin(x^2), x);
```

$$2 \ x \ \cos(x^2)$$

If the expression contains symbolic calls to functions whose derivative is not known, then diff returns itself symbolically:

```
>> diff(x*f(x), x);
```

```
    f(x) + x diff(f(x), x)
```

You may compute higher derivatives via diff(expression,x,x,..). The sequence x,x,.. of identifiers may be generated conveniently via the sequence operator $ (Section 4.5):

```
>> diff(sin(x^2), x, x, x) = diff(sin(x^2), x $ 3);
```

$$- \ 12 \ x \ \sin(x^2) - 8 \ x^3 \ \cos(x^2) =$$

$$- \ 12 \ x \ \sin(x^2) - 8 \ x^3 \ \cos(x^2)$$

You can compute partial derivatives in the same way. Note that Mu-PAD does not assume that mixed partial derivatives of symbolic expressions are symmetric, since mathematically symmetry only holds when the function is sufficiently smooth:

```
>> diff(f(x,y), x, y) - diff(f(x,y), y, x);

   diff(f(x, y), x, y) - diff(f(x, y), y, x)
```

If a mathematical map is represented by a function instead of an expression, then the differential operator D computes the derivative as a function:

```
>> D(sin), D(exp), D(ln), D(sin*cos), D(sin@ln), D(f+g);

             1        2       2   cos@ln
   cos, exp, --, cos   - sin , ------, D(f) + D(g)
             id                   id

>> f := x -> (sin(ln(x))): D(f);

   cos@ln
   ------
     id
```

Here id denotes the identity map $x \mapsto x$. The expression D(f)(x) returns the value of the derivative at a point:

```
>> D(f)(1), D(f)(y^2), D(g)(0);

                      2
            cos(ln(y ))
   cos(0), ----------, D(g)(0)
                2
               y
```

The system converts the prime ' for the derivative to a call of D:

```
>> f'(1), f'(y^2), g'(0);
```

$$\cos(0), \quad \frac{\cos(\ln(y^2))}{y^2}, \quad D(g)(0)$$

For a function with more than one argument, `D([i],f)` is the partial derivative with respect to the ith argument, and `D([i,j,..],f)` is equivalent to `D([i],D([j],(..)))`, for higher partial derivatives.

Exercise 7.1: Consider the function $f : x \to \sin(x)/x$. Compute first the value of f at the point $x = 1.23$, and then the derivative $f'(x)$. Why does:

```
>> f := sin(x)/x: x := 1.23: diff(f, x);
```

not yield the desired result?

Exercise 7.2: De l'Hospital's rule states that

$$\lim_{x \to x_0} \frac{f(x)}{g(x)} = \lim_{x \to x_0} \frac{f'(x)}{g'(x)} = \ldots = \lim_{x \to x_0} \frac{f^{(k-1)}(x)}{g^{(k-1)}(x)} = \frac{f^{(k)}(x_0)}{g^{(k)}(x_0)},$$

if $f(x_0) = g(x_0) = \ldots = f^{(k-1)}(x_0) = g^{(k-1)}(x_0) = 0$ and $g^{(k)}(x_0) \neq 0$.
Compute $\lim_{x \to 0} \dfrac{x^3 \sin(x)}{(1 - \cos(x))^2}$ by applying this rule interactively. Use the function `limit` to check your result.

Exercise 7.3: Determine the first and second order partial derivatives of $f_1(x_1, x_2) = \sin(x_1 x_2)$. Let $x = x(t) = \sin(t)$, $y = y(t) = \cos(t)$, and $f_2(x, y) = x^2 y^2$. Compute the derivative of $f_2(x(t), y(t))$ with respect to t.

7.2 Integration

The function `int` features both definite and indefinite integration:

```
>> int(sin(x), x), int(sin(x), x = 0..PI/2);
```

```
-cos(x), 1
```

If int is unable to compute a result, then it returns itself symbolically. In the following example, the integrand is split into two terms internally. Only one of these has an integral that can be respresented by elementary functions:

```
>> int((x-1)/(x*sqrt(x^3+1)), x);
```

```
  /       3     1/2    3          3     1/2 \
  | (- (x   + 1)    - 1) (1 - (x   + 1)    ) |
ln| -------------------------------------- |
  |                    6                    |
  \                    x                    /
---------------------------------------------- -
                    3
```

```
     /          1            \
  int| - -----------,  x |
     |      3    1/2     |
     \   (x  + 1)        /
```

The function $\mathrm{erf}(x) = \dfrac{2}{\sqrt{\pi}} \displaystyle\int_0^x e^{-t^2}\, dt$ is implemented as a special function in MuPAD:

```
>> int(exp(-a*x^2), x);
```

```
        1/2
  erf(a    x)
  -----------
    / a \1/2
  2 | -- |
    \ PI /
```

Besides the exact computation of definite integrals, MuPAD also provides several numerical methods:

```
>> float(int(exp(-x^2), x = 0..2));
```

```
  0.8820813907
```

In the previous computation, `int` first returns a symbolic result (the erf function), which is then approximated by `float`. If you want to compute numerically from the beginning, then you can suppress the symbolic computation via `int` by using `hold` (Section 5.2):

```
>> float(hold(int)(exp(-x^2), x = 0..2));
```

    0.8820813907

Alternatively, you can use the function `numeric::quadrature` from the `numeric` library:

```
>> numeric::quadrature(exp(-x^2), x = 0..2);
```

    0.8820813907

This function allows you to choose different numerical methods for computing the integral. You find more detailed information with `?numeric::quadrature`. It works in a purely numerical fashion without any symbolic preprocessing of the integrand. The integrand should be smooth and without singularities. Then `numeric::quadrature` is very efficient.

Exercise 7.4: Compute the following integrals:

$$\int_0^{\pi/2} \sin(x)\,\cos(x)\,dx \ , \quad \int_0^1 \frac{dx}{\sqrt{1-x^2}} \ , \quad \int_0^1 x\,\arctan(x)\,dx \ .$$

Use MuPAD to verify the following equality: $\displaystyle\int_{-2}^{-1} \frac{dx}{x} = -\ln(2)$.

Exercise 7.5: Use MuPAD to determine the following indefinite integrals:

$$\int^x \frac{t\,dt}{\sqrt{(2\,a\,t - t^2)^3}} \ , \quad \int^x \sqrt{t^2 - a^2}\,dt \ , \quad \int^x \frac{dt}{t\,\sqrt{1+t^2}} \ .$$

Exercise 7.6: The function **changevar**[1] performes a change of variable in a symbolic integral. Read the corresponding help page. MuPAD version 1.4 cannot compute the integral

$$\int_{-\pi/2}^{\pi/2} \sin(x) \sqrt{1+\sin(x)} \; dx.$$

Assist the system by using the substitution **t=sin(x)**. Compare the value that you get to the numerical result returned by the function **numeric::quadrature**.

---

[1] This function is renamed to **intlib::changevar** in MuPAD versions beyond 1.4.

8. Solving Equations: `solve`

The function `solve` solves systems of equations. This routine can handle various different types of equations. Besides "algebraic" equations, also certain classes of differential and recurrence equations can be solved.

8.1 Polynomial Equations

You can supply an individual equation as first argument to `solve`. The unknown for which you want to solve is the second argument:

```
>> solve(x^2 + x = y/4, x), solve(x^2 + x - y/4 = 0, y);
```

```
{            1/2                   1/2          }
{    (y + 1)                  (y + 1)           }              2
{ - ----------- - 1/2, ----------- - 1/2 }, {4 x + 4 x }
{        2                     2               }
```

In this case, the system returns a set of solutions. If you specify an expression instead of an equation, then `solve` assumes the equation `expression=0`:

```
>> solve(x^2 + x - y/4, y);
```

```
              2
    {4 x + 4 x }
```

For polynomials of higher degree, it is provably impossible to find always a closed form for the solutions by means of radicals etc. In such cases MuPAD uses the `RootOf` object:

```
>> solve(x^7 + x^2 + x, x);
```

```
                     6
  {0, RootOf(x + x  + 1)}
```

Here, the object `RootOf(x+x^6+1)` represents all solutions of the equation `x+x^6+1=0`. You can use `float` to approximate such objects by floating-point numbers. The system internally employs a numerical procedure to determine all (complex) roots of the polynomial:

```
>> map(%, float);
```

```
  {0.0, - 0.7906671888 + 0.3005069203 I,

     - 0.7906671888 - 0.3005069203 I,

     0.9454023333 + 0.6118366937 I,

     0.9454023333 - 0.6118366937 I,

     - 0.1547351444 - 1.038380754 I,

     - 0.1547351444 + 1.038380754 I}
```

If you want to solve a collection of equations for possibly several unknowns, specify both the equations and the unknowns as sets. In the following example, we solve two *linear* equations in three unknowns:

```
>> equations := {x + y + z = 3, x + y = 2}:
>> solution := solve(equations, {x, y, z});
```

```
  {{z = 1, x = 2 - y}}
```

If you solve equations for several variables, then MuPAD returns a set of "solved equations" equivalent to the original system of equations. You can now read off the solutions immediately: the unknown z has the value 1, the unknown y may be arbitrary, and for any given value of z we have $x = 2 - y$. The call to `solve` does not assign the values to the unknowns; x and z are still unknowns. However, the form of the output, as a set of solved equations, is chosen in such a way that you can conveniently use `subs` (Chapter 6) to substitute these values in other objects. For example, you may substitute the solution in the original equations to verify the result:

```
>> subs(equations, op(solution));
```

```
   {2 = 2, 3 = 3}
```

You can use `assign(op(solution))` to assign the solution values to the identifiers x and z.

The output format is different for *nonlinear* polynomial equations in several unknowns:

```
>> equations := {x^2 + y = 1, x - y = 2}:
>> solve(equations, {x, y});
```

```
  { --                   1/2        --
  { |              13            |
  { |   y = x - 2, x = ------ - 1/2  |,
  { --                 2          --

       --                   1/2        -- }
       |              13            |  }
       |   y = x - 2, x = - ------ - 1/2  | }
       --                   2          -- }
```

We first note that MuPAD has found two distinct solutions. Each solution comprises an ordered list of solved equations. The order of the simplified equations is relevant here: the last equation determines x, and the solution for y is expressed symbolically in terms of x. If you use the option BackSubstitution=TRUE, then the system returns a "substituted" form of the solution[1]:

```
>> solutions := solve(equations, {x, y},
                  BackSubstitution = TRUE);
```

```
  { --         1/2             1/2        --
  { |     13                13           |
  { |   y = ----- - 5/2, x = ------ - 1/2  |,
  { --       2                 2          --
```

---

[1] In MuPAD versions beyond 1.4, BackSubstitution=TRUE is the default behavior of `solve`, and you need to pass the option BackSubstitution=FALSE to suppress the substitution.

```
   --              1/2              1/2        -- }
   |              13              13          | }
   |  y = - ------ - 5/2, x = - ------ - 1/2  | }
   --             2                2          -- }
```

Again, you can use **subs** to substitute the solutions in other expressions:

```
>> map(subs(equations, op(solutions, 1)), expand),
   map(subs(equations, op(solutions, 2)), expand);

   {1 = 1, 2 = 2}, {1 = 1, 2 = 2}
```

Often solutions can only be represented by **RootOf** expressions:

```
>> solve({x^3 + x^2 + 2*x = y, y^2 = x^3}, {x, y});

                2   3
   {[y = 2 x + x  + x , x = RootOf(

              2     3    4
      3 x + 5 x + 2 x + x  + 4)], [y = 0, x = 0]}
```

If you use the option **MaxDegree=n**, then **RootOf** expressions for polynomials of degree up to n are replaced by representations in terms of radicals, if this is possible:

```
>> solve({x^3 + x^2 + 2*x = y, y^2 = x^3}, {x, y},
         MaxDegree = 4);

   { --              2   3
   { |  y = 2 x + x  + x , x = ((
   { --

             1/2   1/2              2/3
      9 (5/18 I 3     73    - 785/54)      -

           --                           }
    - ... |, ... , [y = 0, x = 0] }
           --                           }
```

Specifying the unknowns to solve for is optional. If you supply only equations, then **solve** internally uses the system function **indets** to find the symbolic identifiers in the equations and regards all of them as unknowns:

```
>> solve({x + y^2 = 1, x^2 - y = 0});
```

$$\{[x = 1 - y^2 , y = \text{RootOf}(- y - 2 y^2 + y^4 + 1)]\}$$

You should be cautious here, since up to version 1.4 MuPAD also regards system constants such as PI as identifiers:

```
>> solve({x + y^2 = PI, x^2 - y = 1});
```

$$\{[\text{PI} = x - 2 x^2 + x^4 + 1, y = x^2 - 1]\}$$

For polynomial equations, **solve** employs the tools from the **groebner** library.

You can use the function **float** to find *numerical* solutions. However, with a statement of the form

```
>> float(solve(equations, unknowns)):
```

solve first tries to solve the equations symbolically, and **float** then handles the result returned by **solve**. If you want to compute in a purely numerical way, then you can use **hold** (Section 5.2) to avoid symbolic preprocessing:

```
>> float(hold(solve)({x^3 + x^2 + x + x = y, y^2 = x^3},
                      {x, y}));
```

$$\{[y = 0.0, x = 0.0], [y = 2.453618103 - 0.1141849396 \text{ I},$$

$$x = - 0.8609295554 + 1.604034314 \text{ I}], [$$

$$y = 2.453618103 + 0.1141849396 \text{ I},$$

$$x = - 0.8609295554 - 1.604034314 \text{ I}], [$$

    y = - 0.9536181035 + 0.6454386764 I,

    x = - 0.1390704445 + 1.089777135 I], [

    y = - 0.9536181035 - 0.6454386764 I,

    x = - 0.1390704445 - 1.089777135 I]}

Furthermore, the library **numeric** provides functions such as **newton** or **fsolve**[2] for numerical equation solving. The help system gives details about these routines: **?numeric::newton** etc.

Exercise 8.1: Compute the general solution of the system of linear equations
$$
\begin{aligned}
a + b + c + d + e &= 1\,, \\
a + 2b + 3c + 4d + 5e &= 2\,, \\
a - 2b - 3c - 4d - 5e &= 2\,, \\
a - b - c - d - e &= 3\,.
\end{aligned}
$$

How many free parameters does the solution have?

8.2 General Equations

MuPAD's **solve** can handle a variety of (non-polynomial) equations. For example, the equation $\exp(x) = 8$ has infinitely many solutions of the form $\ln(8) + k\,2\,\pi\,I$ for $k = 0, \pm 1, \pm 2, \ldots$ in the complex plane:

```
>> solve(exp(x) = 8, x);
```

    {ln(8), (- 2 I) PI + ln(8), 2 I PI + ln(8), ...}

The data type of the returned result is a so-called "discrete set"[3] which is capable of representing infinitely many elements:

---

[2] In MuPAD versions beyond 1.4, **numeric::newton** and **numeric::fsolve** are replaced by **numeric::solve** and **numeric::realroots**, respectively.

[3] In MuPAD versions beyond 1.4, the result is of a different data type; see Section 8.4.

```
>> solutions := op(%): domtype(solutions);
```

    Dom::DiscreteSet

This set has infinitely many operands, for example:

```
>> op(solutions, 78), op(solutions, 79);
```

    ln(8) - 78 I PI, 78 I PI + ln(8)

The equation `exp(x)=sin(x)` also has infinitely many solutions, which, however, MuPAD cannot represent exactly. In this case it returns the call to `solve` symbolically:

```
>> solutions := solve(exp(x) = sin(x), x);
```

    solve(exp(x) = sin(x), x)

Warning: In contrast to polynomial equations, the numerical solver computes at most one solution of a non-polynomial equation:

```
>> float(solutions);
```

    {-3.183063012}

However, you can specify a search range for a particular numerical solution:

```
>> float(hold(solve)(exp(x) = sin(x), x = -10..-9));
```

    {-9.424858654}

8.3 Differential and Recurrence Equations

The function `ode` defines an ordinary differential equation. Such an object has two components: an equation and the function to solve for.

```
>> diffEquation := ode(y'(x) = y(x)^2, y(x));
```

```
                          2
   ode({diff(y(x), x) = y(x) }, y(x))
```

The following call to `solve` finds the general solution containing an arbitrary constant C_1:

```
>> solve(diffEquation, y(x));

   {    1     }
   { -------- }
   { - C1 - x }
```

Differential equations of higher order can be handled as well:

```
>> solve(ode(y''(x) = y(x), y(x)), y(x));

   {C2 exp(x) + C3 exp(-x)}
```

You can specify initial conditions in form of a set when calling `ode`:

```
>> diffEquation :=
     ode({y''(x) = y(x), y(0) = 1, y'(0) = 0}, y(x)):
```

MuPAD now adjusts the free constants in the general solution according to the initial conditions:

```
>> solve(diffEquation);

   { exp(x)    exp(-x) }
   { ------ + ------- }
   {   2         2     }
```

Here we have supplied only the differential equation as argument to `solve` and not the function for which we want to solve. Indeed, specifying the second argument is optional: `solve` automatically finds the object to solve for, since it is part of the data structure generated by `ode`.

You can specify systems of equations with several functions in form of a set:

```
>> solve(ode({y'(x) = y(x) + z(x), z'(x) = y(x)},
            {y(x), z(x)}));
```

```
{                    /          1/2 \        /          1/2 \
{                    | x     x  5   |        | x     x  5   |
{              C6 exp| -  -  -----  |   C7 exp| -  +  -----  |
{                    \ 2      2    /         \ 2      2    /
{ y(x) =  ---------------------------  +  ---------------------------
{                     2                              2
```

```
              /          1/2 \
     1/2      | x     x  5   |
   C6 5    exp| -  -  -----  |
              \ 2      2    /
  -  -----------------------------  +
              2
```

```
              /          1/2 \
     1/2      | x     x  5   |
   C7 5    exp| -  +  -----  |
              \ 2      2    /
  -----------------------------,
              2
```

```
                                                                    }
              /          1/2 \        /              1/2 \ }
              | x     x  5   |        | x       x  5   | }
  z(x) = C6 exp| -  -  -----  |  +  C7 exp| -  +  -----  | }
              \ 2      2    /        \ 2       2    / }
                                                                    }
```

The function numeric::odesolve of the numeric library solves the ordinary differential equation $Y'(x) = f(x, Y(x))$ with the initial condition $Y(x_0) = Y_0$ numerically. You must supply the right hand side of the differential equation as a function $f(x, Y)$ of two arguments, where x is a scalar and Y is a vector. If you combine the components y and z in the previous example to a vector $Y = (y, z)$, then you can define the right hand side of the equation

$$\frac{d}{dx} Y = \frac{d}{dx} \begin{pmatrix} y \\ z \end{pmatrix} = \begin{pmatrix} y+z \\ y \end{pmatrix} = \begin{pmatrix} Y[1] + Y[2] \\ Y[1] \end{pmatrix} =: f(x, Y)$$

in the form

```
>> f := (x, Y) -> ([Y[1] + Y[2], Y[1]]):
```

Note that $f(x, y)$ must be a vector. Here this is realized by means of a list containing the components on the right hand side of the differential equation. The call

```
>> numeric::odesolve(0..1, f, [1, 1]);
```

```
   [5.812568463, 3.798245729]
```

integrates the system of differential equations with the initial values $Y(0) = (y(0), z(0)) = (1, 1)$, which are specified as a list, over the interval $x \in [0, 1]$. It returns the numerical solution vector $Y(1) = (y(1), z(1))$.

Recurrence equations are equations for functions depending on a discrete parameter (an "index"). You can generate such an object with the function **rec**, whose arguments are an equation, the function to be determined, and, optionally, a set of initial conditions:

```
>> equation := rec(x(n + 2) = x(n + 1) + 2*x(n), x(n)):
>> solve(equation, x(n));
```

$$\{a1\ (-1)^n + a2\ 2^n\}$$

The second argument `x(n)` is again optional, since **solve** automatically recognizes the unknown function as the second component of the data structure generated by **rec**. The general solution contains two arbitrary constants a_1, a_2, which are suitably adjusted when you specify initial conditions:

```
>> solve(rec(x(n + 2) = 2*x(n) + x(n + 1), x(n),
            {x(0) = 1}), x(n));
```

$$\{a4\ 2^n + (-1)^n\ (1 - a4)\}$$

```
>> solve(rec(x(n + 2) = 2*x(n) + x(n + 1), x(n),
            {x(0) = 1, x(1) = 1}));
```

```
{     n        n }
{  (-1)       2 2  }
{  ----- +  ----  }
{    3        3   }
```

Exercise 8.2: Check the numerical solutions $y(1) = 5.812\ldots$ and $z(1) = 3.798\ldots$ of the system of differential equations

$$y'(x) = y(x) + z(x) \ , \ \ z'(x) = y(x)$$

computed above by substituting the initial values $y(0) = 1$, $z(0) = 1$ in the general symbolic solution, determining the values for the free constants, and evaluating the symbolic solution at $x = 1$.

Exercise 8.3:

1) Compute the general solution $y(x)$ of the differential equation $y' = y^2/x$.
2) Determine the solution $y(x)$ for each of the following initial value problems:

   a) $y' - y \sin(x) = 0$, $y'(1) = 1$,   b) $2y' + \dfrac{y}{x} = 0$, $y'(1) = \pi$.

3) Find the general solution of the following system of ordinary differential equations in $x(t), y(t), z(t)$:

$$x' = -3\,y\,z \ , \ \ y' = 3\,x\,z \ , \ \ z' = -x\,y .$$

Exercise 8.4: The Fibonacci numbers are defined by the recurrence $F_n = F_{n-1} + F_{n-2}$ with the initial values $F_0 = 0$, $F_1 = 1$. Use `solve` to find an explicit representation for F_n.

8.4 `solve` in MuPAD Versions Beyond 1.4

The behavior of the function `solve` in MuPAD versions beyond 1.4 differs substantially from the behavior in version 1.4. This section summarizes the main changes.

The output format for polynomial equations changes slightly, but it is also more consistent. The general rule is as follows:

A call of the form `solve(equation,unknown)`, where `equation` is a single equation (or polynomial) and `unknown` is an identifier, returns a set of MuPAD objects representing numbers. All other forms of `solve` for (one or several) polynomial equations return a set of lists of equations.

Here are some examples:

```
>> solve(x^2 - 3*x + 2 = 0, x), solve(x^2 - 3*x + 2, x);

   {1, 2}, {1, 2}

>> solve(x^2 - 3*x + 2 = 0), solve(x^2 - 3*x + 2);

   {[x = 1], [x = 2]}, {[x = 1], [x = 2]}

>> solve({x^2 - 3*x + 2 = 0}, x),
   solve({x^2 - 3*x + 2}, x);

   {[x = 1], [x = 2]}, {[x = 1], [x = 2]}

>> solve({x^2 - 3*x + y = 0, y - 2*x = 0}, {x, y});

   {[x = 0, y = 0], [x = 1, y = 2]}

>> solve({x^2 - 3*x + y = 0, y - 2*x = 0});

   {[x = 0, y = 0], [x = 1, y = 2]}
```

By default, `solve` tries to find all *complex* solutions of the given equation(s). If you want to find only the real solutions of a single equation, use the option `Domain=Dom::Real`:

```
>> solve(x^3 + x = 0, x);

   {0, - I, I}

>> solve(x^3 + x = 0, x, Domain = Dom::Real);

   {0}
```

By specifying `Domain=Dom::Rational` or `Domain=Dom::Integer`, you obtain only rational or integral solutions, respectively. If all complex numbers satisfy a given equation, then `solve` returns[4] `C_`. This is a MuPAD object that represents the mathematical set of complex numbers:

```
>> solve(sin(x)^2 + cos(x)^2 = 1, x);

   C_

>> domtype(%);

   solvelib::BasicSet
```

There are four such "basic sets" in MuPAD versions beyond 1.4: the integers `Z_`, the rational numbers `Q_`, the real numbers `R_`, and the complex numbers `C_`.

For more general equations, MuPAD versions beyond 1.4 provide a new data type: image sets, of domain type `Dom::ImageSet`. They are to represent mathematical sets of the form $\{f(x) \mid x \in A\}$, where A is some other set. For instance, the first example in Section 8.2 works as follows:

```
>> S := solve(exp(x) = 8, x);

   { 2*I*PI*X1 + ln(8) |  X1 in Z_ }

>> domtype(S);

   Dom::ImageSet
```

If you omit the variable to solve for, then the system returns a logical formula, using the new MuPAD operator `in`:

```
>> solve(exp(x) = 8);

   x in { 2*I*PI*X2 + ln(8) |  X2 in Z_ }
```

---

[4] In MuPAD versions up to 1.4, the system returns a set with the variable to solve for instead.

You can use `map` to apply a function to an image set:

```
>> map(S, Im);
```

```
{ 2*X2*PI | X2 in Z_ }
```

The function `is` (Section 9.3) can handle objects of this type:

```
>> S := solve(sin(PI*x/2) = 0, x);
```

```
{ 2*X3 | X3 in Z_ }
```

```
>> is(1 in S), is(4 in S);
```

```
FALSE, TRUE
```

The function `solve` can also handle inequalities. It then returns an interval or a union of intervals, of domain type `Dom::Interval`:

```
>> solve(x^2 < 1, x);
```

```
]-1, 1[
```

```
>> domtype(%);
```

```
Dom::Interval
```

```
>> S := solve(x^2 >= 1, x);
```

```
[1, infinity[ union ]-infinity, -1]
```

```
>> is(-2 in S), is(0 in S);
```

```
TRUE, FALSE
```

MuPAD versions beyond 1.4 provide a special data type for the solution of a parametric equation: `piecewise`. For example, the set of solutions $x \in \mathbb{C}$ of the equation $(ax^2 - 4)(x - b) = 0$ depends on the value of the parameter[5] a:

---

[5] In MuPAD versions up to 1.4, `solve` returns the solution that is valid if $a \neq 0$.

```
>> p:= solve((a*x^2 - 4)*(x - b), x);
```

```
  {b}                               if a = 0
  {b, -2/a^(1/2), 2/a^(1/2)}  if a <> 0
```

```
>> domtype(p);
```

```
  piecewise
```

The function **map** applies a function to all branches of a **piecewise** object:

```
>> map(p, _power, 2);
```

```
  {b^2}              if a = 0
  {b^2, 4/a}    if a <> 0
```

After the following substitution, the **piecewise** object is simplified to a set:

```
>> eval(subs(%, [a = 4, b = 2]));
```

```
  {1, 4}
```

9. Manipulating Expressions

When evaluating objects, MuPAD automatically performs a variety of simplifications. For example, arithmetic operations between integers are executed or $\exp(\ln(x))$ is simplified to x. Other mathematically possible simplifications such as $\sin(x)^2 + \cos(x)^2 = 1$, $\ln(\exp(x)) = x$, $(x^2 - 1)/(x - 1) = x + 1$, or $\sqrt{x^2} = x$ do not happen automatically. The reason is for one that such rules are not universally valid: for example, $\sqrt{x^2} = x$ is wrong for $x = -2$. Other simplifications such as $\sin(x)^2 + \cos(x)^2 = 1$ are valid universally, but there would be a significant loss of efficiency, if MuPAD would always scan expressions for the occurrence of sin and cos terms.

Moreover, it is not clear in general which of several mathematically equivalent representations is the most appropriate. For example, it might be reasonable to replace an expression such as $\sin(x)$ by its complex exponential representation

$$\sin(x) = -\frac{I}{2} \exp(x\,I) + \frac{I}{2} \exp(-x\,I).$$

In such a situation, you can control the manipulation and simplification of expressions by explicitly applying appropriate system functions. MuPAD provides the following functions, which we have partly discussed in Section 2.3:

`collect`	:	collecting coefficients
`combine`	:	combining subexpressions
`expand`	:	expansion
`Factor`	:	factorization
`normal`	:	normalization of rational expressions
`partfrac`	:	partial fraction decomposition
`radsimp`	:	simplification of radicals
`rectform`	:	Cartesian representation of complex values
`rewrite`	:	applying mathematical identities
`simplify`	:	universal simplifier

9.1 Transforming Expressions

If you enter the command `collect(expression,unknown)`, the system regards the expression as a polynomial in the specified unknown(s) and groups the coefficients of equal powers together:

```
>> x^2 + a*x + sqrt(2)*x + b*x^2 + sin(x) + a*sin(x):
>> collect(%, x);
```

$$
sin(x) + a\ sin(x) + x^2\ (b + 1) + x\ (a + 2^{1/2})
$$

You can specify several "unknowns", which may themselves be expressions, as a list:

```
>> collect(%, [x, sin(x)]);
```

$$
sin(x)\ (a + 1) + x^2\ (b + 1) + x\ (a + 2^{1/2})
$$

The function `combine(expression,option)` combines subexpressions by using mathematical identities between functions given by `option`. The possible options are `_power`, `atan`, `exp`, `ln`, `sincos`, `sinhcosh`, `sqrt`.

When you specify the option `_power`, then MuPAD employs the identity $a^c\,b^c = (a\,b)^c$ for powers:

```
>> f := a^x*3^y/2^x/9^y: f := combine(f, _power);
```

$$
(1/3)^y\ \left(\frac{a}{2}\right)^x
$$

The inverse $atan^1$ of the tangent function satisfies the following identity:

```
>> f := atan(x) + atan(y): f = combine(f, atan);
```

---

[1] The functions asin, acos, atan, asinh, etc., are renamed to arcsin, arccos, arctan, arcsinh, etc., in MuPAD versions beyond 1.4.

$$\text{atan}(x) + \text{atan}(y) = \text{atan}\left(\frac{x + y}{1 - x\,y}\right)$$

For the exponential function, we have $\exp(x)\exp(y) = \exp(x+y)$ and $\exp(x)^y = \exp(x\,y)$:

```
>> combine(exp(x)*exp(y)^2/exp(-z), exp);

  exp(x + 2 y + z)
```

With certain assumptions about x, y, the logarithm satisfies the rules $\ln(x) + \ln(y) = \ln(x\,y)$ and $x\,\ln(y) = \ln(y^x)$:

```
>> combine(ln(x) + ln(2) + 3*ln(3/2), ln);

    / 16 x \
  ln| ---- |
    \  27  /
```

The trigonometric functions satisfy a variety of identities that the system employs to combine products:

```
>> combine(sin(x)*cos(y), sincos),
   combine(sin(x)^2, sincos);

  sin(x + y)    sin(x - y)           cos(2 x)
  ---------- + ----------- , 1/2 - --------
      2             2                  2
```

Similar rules are applied to the hyperbolic functions:

```
>> combine(sinh(x)*cosh(y), sinhcosh),
   combine(sinh(x)^2, sinhcosh);

  sinh(x + y)    sinh(y - x)   cosh(2 x)
  ----------- + ----------- , --------- - 1/2
       2             2            2
```

Finally, the option sqrt combines expressions with square roots:

```
>> combine(sqrt(6)*sqrt(7)*sqrt(x), sqrt);
```

```
        1/2
  (42 x)
```

The function **expand** applies the identities used by **combine** in the reverse direction: it transforms special function calls with composite arguments to sums or products of function calls with simpler arguments via "addition theorems":

```
>> expand(x^(y + z)), expand(exp(x + y - z + 4)),
   expand(ln(2*PI*x*y));
```

```
  y z   exp(x) exp(y) exp(4)
 x  x , ----------------------, ln(PI) + ln(x y) + ln(2)
               exp(z)
```

```
>> expand(sin(x + y)), expand(cosh(x + y));
```

```
  cos(x) sin(y) + cos(y) sin(x),

      cosh(x) cosh(y) + sinh(x) sinh(y)
```

```
>> expand(sqrt(42*x*y));
```

```
        1/2    1/2
  (x y)     42
```

Here the system does not perform some "expansions" such as $\ln(x\,y) = \ln(x) + \ln(y)$, since such an identity holds only under additional assumptions (for example, for positive real x and y).

The most frequent use of **expand** is for transforming a product of sums into a sum of products:

```
>> expand((x + y)^2*(x - y)^2);
```

```
   4    4      2 2
  x  + y  - 2 x  y
```

This works recursively for all subexpressions:

```
>> expand((x - y)*(x + y)*sin(exp(x + y + z)));
```

$$x^2 \; \sin(\exp(x)\; \exp(y)\; \exp(z)) - $$

$$y^2 \; \sin(\exp(x)\; \exp(y)\; \exp(z))$$

You can supply expressions as additional arguments to **expand**. These subexpressions are *not* expanded:

```
>> expand((x - y)*(x + y)*sin(exp(x + y + z)),
          x - y, x + y + z);
```

```
  x sin(exp(x + y + z)) (x - y) +

    y sin(exp(x + y + z)) (x - y)
```

The function **Factor**[2] factors polynomials and expressions:

```
>> Factor(x^3 + 3*x^2 + 3*x + 1);
```

$$(x + 1)^3$$

Here the system factors "over the rational numbers": it looks for polynomial factors with rational number coefficients. In effect, MuPAD does not return the factorization[3] $x^2 - 2 = (x - \sqrt{2})\,(x + \sqrt{2})$:

```
>> Factor(x^2 - 2);
```

----

[2] The function **Factor** is replaced by **factor** in MuPAD versions beyond 1.4; see the footnote on page 135.

[3] There is, however, a possibility to factor over other rings. For that purpose, you must transform the expression into a polynomial over the corresponding coefficient ring. For example, if we choose the extension field of the rational numbers with $Z = \sqrt{2}$ that we have already considered in Section 4.14

```
>> K := Dom::AlgebraicExtension(Dom::Rational, Z^2 = 2, Z):
```

then we can factor the polynomial

```
>> p := poly(x^2 - 2, [x], K):
```

over the ring K:

```
  2
x   - 2
```

For sums of rational expressions, `Factor` first computes a common denominator and then factors both the numerator and the denominator:

```
>> f := (x^3 + 3*y^2)/(x^2 - y^2) + 3: f = Factor(f);
```

```
  3     2              2
 x  + 3 y            x  (x + 3)
 --------- + 3 =   ---------------
   2    2          (x + y) (x - y)
  x  - y
```

MuPAD can factor not only polynomials and rational functions. For more general expressions, the system internally replaces subexpressions such as symbolic function calls by identifiers, factors the corresponding polynomial or rational function, and resubstitutes the temporary identifiers:

```
>> Factor((exp(x)^2 - 1)/(sin(x)^2 - cos(x)^2));
```

```
      (exp(x) - 1) (exp(x) + 1)
 -------------------------------------
 (cos(x) + sin(x)) (sin(x) - cos(x))
```

The function `normal` computes a "normal form" for rational expressions. Like `Factor`, it first computes a common denominator for sums of rational expressions, but it then expands numerator and denominator instead of factoring them:

```
>> f := ((x + 6)^2 - 17)/(x - 1)/(x + 1) + 1:
   f, Factor(f), normal(f);
```

```
>> Factor(p);
   poly(x - Z, [x], Dom::AlgebraicExtension(Dom::Rational,
      2
      Z - 2 = 0, Z)) poly(x + Z, [x],
                                        2
   Dom::AlgebraicExtension(Dom::Rational, Z - 2 = 0, Z))
```

$$\frac{(x + 6)^2 - 17}{(x - 1)\,(x + 1)} + 1, \quad \frac{2\,(x + 3)^2}{(x - 1)\,(x + 1)}, \quad \frac{12\,x + 2\,x^2 + 18}{x^2 - 1}$$

Nevertheless, **normal** cancels common factors in numerator and denominator:

```
>> f := x^2/(x + y) - y^2/(x + y): f = normal(f);
```

$$\frac{x^2}{x + y} - \frac{y^2}{x + y} = x - y$$

Like **Factor**, **normal** can handle arbitrary expressions:

```
>> f := (exp(x)^2-exp(y)^2)/(exp(x)^3 - exp(y)^3):
>> f = normal(f);
```

$$\frac{\exp(x)^2 - \exp(y)^2}{\exp(x)^3 - \exp(y)^3} = \frac{\exp(x) + \exp(y)}{\exp(x)\,\exp(y) + \exp(x)^2 + \exp(y)^2}$$

The function **partfrac** decomposes a rational expression into a polynomial part plus a sum of rational terms whose numerator degree is smaller degree than the corresponding denominator degree (partial fraction decomposition):

```
>> f := x^2/(x^2 - 1): f = partfrac(f, x);
```

$$\frac{x^2}{x^2 - 1} = \frac{1}{2\,(x - 1)} - \frac{1}{2\,(x + 1)} + 1$$

The denominators of the terms are the factors that MuPAD finds when factoring the common denominator:

```
>> denominator := x^5 + x^4 - 7*x^3 - 11*x^2 - 8*x - 12:
>> Factor(denominator);
```

```
        2              2
(x + 2)   (x - 3) (x  + 1)
```

```
>> partfrac(1/denominator, x);
```

```
                                           9 x
                                           --- - 13/250
                                           250
     1              1            1              
------------  -  ----------  -  -----------  +  ------------
250 (x - 3)     25 (x + 2)                 2       2
                                 25 (x + 2)       x  + 1
```

Another function for manipulating expressions is **rewrite**. It employs identities to eliminate certain functions completely from an expression and replaces them by different functions. For example, you can always express sin and cos by tan with the half argument. The trigonometric functions are also related to the complex exponential function:

$$\sin(x) = \frac{2\tan(x/2)}{1+\tan(x/2)^2} , \quad \cos(x) = \frac{1-\tan(x/2)^2}{1+\tan(x/2)^2} ,$$

$$\sin(x) = -\frac{I}{2} \exp(I\,x) + \frac{I}{2} \exp(-I\,x) ,$$

$$\cos(x) = \frac{1}{2} \exp(I\,x) + \frac{1}{2} \exp(-I\,x) .$$

You can express the hyperbolic functions and their inverse functions in terms of the exponential function and the logarithm:

$$\sinh(x) = \frac{\exp(x) - \exp(-x)}{2} , \quad \cosh(x) = \frac{\exp(x) + \exp(-x)}{2} ,$$

$$\operatorname{asinh}(x) = \ln(x + \sqrt{x^2 + 1}) , \quad \operatorname{acosh}(x) = \ln(x + \sqrt{x^2 - 1}) .$$

A call of the form **rewrite(expression,option)** employs these identities. The following rules are implemented in MuPAD:

option	:	function(s)	→ target
diff	:	differential operator D	→ diff
exp	:	trigonometric and hyperbolic functions	→ exp
fact	:	Γ function gamma	→ fact
gamma	:	factorial fact	→ gamma
heaviside	:	sign	→ heaviside
ln	:	inverse trigonometric and inverse hyperbolic functions	→ ln
sign	:	step function heaviside	→ sign
sincos	:	exponential function exp	→ sin, cos
sinhcosh	:	exponential function exp	→ sinh, cosh
tan	:	sin and cos	→ tan

```
>> rewrite(D(D(u))(x), diff);

  diff(u(x), x, x)

>> rewrite(sin(x)/cos(x), exp) = rewrite(tan(x), exp);


                                             2
1/2 I exp(-I x) - 1/2 I exp(I x)      I exp(I x)  - I
-------------------------------- = - ----------------
       exp(-I x)    exp(I x)                2
       --------- + --------             exp(I x)  + 1
           2          2

>> rewrite(asinh(x) - acosh(x), ln);

           2    1/2            2    1/2
  ln(x + (x  + 1)   ) - ln(x + (x  - 1)   )
```

For expressions representing complex *numbers*, you can easily compute real and imaginary parts by using Re and Im:

```
>> z := 2 + 3*I: Re(z), Im(z);

  2, 3
```

```
>> z := sin(2*I) - ln(-1): Re(z), Im(z);

   0, - PI + sinh(2)
```

When an expression contains symbolic identifiers, then MuPAD assumes all such unknowns to be complex values, and Re and Im are returned symbolically:

```
>> Re(a*b + I), Im(a*b + I);

   Re(a b), Im(a b) + 1
```

In such a case, you can use the function rectform (short for: rectangular form) to decompose the expression into real and imaginary part. The name of this function is derived from the fact that it computes the coordinates of the usual rectangular (Cartesian) coordinate system. MuPAD decomposes the symbols contained in the expression into their real and imaginary parts and expresses the final result accordingly:

```
>> rectform(a*b + I);

   (- Im(a) Im(b) + Re(a) Re(b) ) +

      (Im(a) Re(b) + Im(b) Re(a) + 1) I

>> rectform(exp(x));

   cos(Im(x)) exp(Re(x)) + (sin(Im(x)) exp(Re(x))) I
```

Again, you can extract the real and imaginary parts of the result with Re and Im, respectively:

```
>> Re(%), Im(%);

   cos(Im(x)) exp(Re(x)), sin(Im(x)) exp(Re(x))
```

As a basic principle, rectform regards all symbolic identifiers as representatives of complex numbers. However, you can specify a set of identifiers assumed to be real. They are passed to rectform as a second argument:

```
>> rectform(a*b + I, {a});
```

```
   a Re(b) + (a Im(b) + 1) I
```

Alternatively, `rectform::globalReal(set)`[4] fixes a global set of identifiers that all following calls to `rectform` regard to be real:

```
>> rectform::globalReal({a, x, y}):
>> z := rectform(a*x + y + b);
```

```
   (y + a x + Re(b)) + Im(b) I
```

```
>> rectform(exp(x*I));
```

```
   cos(x) + sin(x) I
```

Moreover, you can use the `assume` command described in Section 9.3:

```
>> assume(a, Type::RealNum): rectform(a*X);
```

```
   a Re(X) + (a Im(X)) I
```

The results of `rectform` have a special data type:

```
>> domtype(z);
```

```
   rectform
```

You can use the function `expr` to convert such an object to a "normal" MuPAD expression of domain type `DOM_EXPR`:

```
>> expr(z);
```

```
   y + a x + I Im(b) + Re(b)
```

We recommend that you apply the function `rectform` only to expressions containing symbolic identifiers. For expressions without such identifiers, `Re` and `Im` return the decomposition into real and imaginary part much faster.

---

[4] In MuPAD versions beyond 1.4, calls to `rectform` with two arguments and `rectform::globalReal` are no longer available; use `assume` instead.

9.2 Simplifying Expressions

In some cases, a transformation leads to a more simple expression:

```
>> f := 2^x*3^x/8^x/9^x: f = combine(f, _power);
```

```
   x  x
  2  3            x
  ----- = (1/12)
   x  x
  8  9
```

```
>> f := x/(x + y) + y/(x + y): f = normal(f);
```

```
    x       y
  ----- + ----- = 1
  x + y   x + y
```

To this end, however, you must inspect the expression and decide yourself which function to use for simplification. However, there is a tool
for applying various simplification algorithms to an expression *auto-*
matically: the function `simplify`. This is a universal simplifier which
MuPAD uses to achieve a representation of an expression that is as
simple as possible:

```
>> f := 2^x*3^x/8^x/9^x: f = simplify(f);
```

```
   x  x
  2  3            x
  ----- = (1/12)
   x  x
  8  9
```

```
>> f := (1 + (sin(x)^2 + cos(x)^2)^2)/sin(x):
>> f = simplify(f);
```

```
        2          2 2
  (cos(x)  + sin(x) )  + 1        2
  ------------------------- = ------
           sin(x)             sin(x)
```

```
>> f := x/(x + y) + y/(x + y) - sin(x)^2 - cos(x)^2:
>> f = simplify(f);
```

$$\frac{x}{x + y} - \cos(x)^2 - \sin(x)^2 + \frac{y}{x + y} = 0$$

```
>> f := (exp(x) - 1)/(exp(x/2) + 1): f = simplify(f);
```

$$\frac{\exp(x) - 1}{\exp\left(\frac{x}{2}\right) + 1} = \exp\left(\frac{x}{2}\right) - 1$$

```
>> f := sqrt(997) - (997^3)^(1/6): f = simplify(f);
```

$$997^{1/2} - 991026973^{1/6} = 0$$

However, the results are not always optimal:

```
>> f := sqrt(4 + 2*sqrt(3)): f = simplify(f);
```

$$2^{1/2}(3^{1/2} + 2)^{1/2} = (2 \cdot 3^{1/2} + 4)^{1/2}$$

You can control the simplification process by supplying additional arguments. As in the case of combine, you can request particular simplifications by means of options. For example, you can tell the simplifier explicitly to simplify expressions containing square roots:

```
>> f = simplify(f, sqrt);
```

$$2^{1/2}(3^{1/2} + 2)^{1/2} = 3^{1/2} + 1$$

The possible options are exp, ln, cos, sin, sqrt, logic, and relation. Internally, simplify then confines itself to those simplification rules that are valid for the function given as option. The options logic and `relation` are for simplifying logical expressions and equations and inequalities, respectively (see also the corresponding help page: `?simplify`).

Instead of `simplify(expression,sqrt)` you may also use the function `radsimp` to simplify numerical expressions containing square roots or other radicals:

```
>> f = radsimp(f);
```

```
   1/2   1/2      1/2      1/2
  2     (3    + 2)     = 3     + 1
```

```
>> f := 2^(1/4)*2 + 2^(3/4) - sqrt(8 + 6*2^(1/2)):
>> f = radsimp(f);
```

```
     1/4    3/4     1/2      1/2      1/2
  2 2    + 2    - 2     (3 2     + 4)     = 0
```

In many cases using `simplify` without options is appropriate. However, such a call is often very time consuming, since the simplification algorithm is quite complex. It may be sensible to specify additional options to save computing time, since then simplifications are performed only for special functions.

Exercise 9.1: It is possible to rewrite products of trigonometric functions as linear combinations of sin and cos terms whose arguments are integral multiplies of the original arguments (Fourier expansion). Find constants a, b, c, d, and e such that

$$\cos(x)^2 + \sin(x)\,\cos(x)$$
$$= a + b\,\sin(x) + c\,\cos(x) + d\,\sin(2\,x) + e\,\cos(2\,x)$$

holds.

Exercise 9.2: Use MuPAD to prove the following identities:

1) $$\frac{\cos(5\,x)}{\sin(2\,x)\cos^2(x)} = -5\,\sin(x) + \frac{\cos^2(x)}{2\,\sin(x)} + \frac{5\,\sin^3(x)}{2\,\cos^2(x)}\,,$$

2) $$\frac{\sin^2(x) - e^{2\,x}}{\sin^2(x) + 2\,\sin(x)\,e^x + e^{2\,x}} = \frac{\sin(x) - e^x}{\sin(x) + e^x}\,,$$

3) $\dfrac{\sin(2\,x) - 5\,\sin(x)\,\cos(x)}{\sin(x)\,(1 + \tan^2(x))} = -\dfrac{9\,\cos(x)}{4} - \dfrac{3\,\cos(3\,x)}{4}$,

4) $\sqrt{14 + 3\,\sqrt{3 + 2\,\sqrt{5 - 12\,\sqrt{3 - 2\,\sqrt{2}}}}} = \sqrt{2} + 3$.

Exercise 9.3: MuPAD computes the following integral for f:

```
>> f := sqrt(sin(x) + 1): int(%, x);

                              1/2
    2 (sin(x) - 1) (sin(x) + 1)
   -------------------------------
             cos(x)
```

Its derivative is not literally identical to the integrand:

```
>> diff(%, x);

                 1/2        sin(x) - 1
    2 (sin(x) + 1)     + --------------- +
                                     1/2
                         (sin(x) + 1)

                                         1/2
      2 sin(x) (sin(x) - 1) (sin(x) + 1)
     -------------------------------------
                        2
                   cos(x)
```

Simplify this expression.

9.3 Assumptions About Symbolic Identifiers

MuPAD performs transformations or simplifications for objects containing symbolic identifiers only if the corresponding rules apply in the entire complex plane. However, some familiar rules for computing with real numbers are not generally valid for complex numbers. For example, the square root and the logarithm are multi-valued complex functions, and the MuPAD functions internally make certain assumptions about the branch cuts:

transformation of	to	generally valid only for
$\ln(e^x)$	x	real x
$\ln(x^n)$	$n\ln(x)$	real $x > 0$
$\ln(x\,y)$	$\ln(x) + \ln(y)$	real $x > 0$ or real $y > 0$
$\sqrt{x^2}$	$\text{sign}(x)\,x$	real x
$e^{x/2}$	$(e^x)^{1/2}$	real x

You can use the function **assume** to tell the system functions such as **expand**, **simplify**, **limit**, **solve**, and **int** that they may make certain assumptions about the meaning of certain identifiers. The corresponding feature is still under development and provides its full functionality in MuPAD versions beyond 1.4. For that reason, we only demonstrate some simple examples here. You find more information on the corresponding help page: ?assume.

You can use a type specifier (Chapter 15) to tell MuPAD that a symbolic identifier represents only values corresponding to the mathematical meaning of the type. For example, the commands[5]

```
>> assume(x, Type::RealNum): assume(y, Type::RealNum):
```

restrict x and y to be real numbers. Now **simplify** can apply additional rules:

```
>> simplify(ln(exp(x))), simplify(sqrt(x^2));
```

```
   x, x sign(x)
```

One of the commands **assume(x>0)** or

```
>> assume(x, Type::Positive):
```

restricts x to positive real numbers, and we have

---

[5] In MuPAD versions beyond 1.4, Type::RealNum is renamed to Type::Real.

```
>> simplify(ln(x^n)), simplify(ln(x*y) - ln(x) - ln(y)),
   simplify(sqrt(x^2));
```

```
   n ln(x), 0, x
```

Transformations and simplifications with constants are always executed without additional assumptions since their mathematical meaning is known:

```
>> expand(ln(2*PI*z)), sqrt((2*PI*z)^2);
```

$$\ln(z) + \ln(PI) + \ln(2), \quad 2 \; PI \; (z^2)^{1/2}$$

The function is checks whether a MuPAD object has a certain mathematical property:

```
>> is(1, Type::Integer), is(PI + 1, Type::RealNum);
```

```
   TRUE, TRUE
```

In addition, is takes into account the assumptions about identifiers set by assume:

```
>> unassign(x): is(x, Type::Integer);
```

```
   UNKNOWN
```

```
>> assume(x, Type::Integer):
>> is(x, Type::Integer), is(x, Type::RealNum);
```

```
   TRUE, TRUE
```

In contrast, the function **testtype** presented in Section 15.1 checks the *technical* type of a MuPAD object:

```
>> testtype(x, Type::Integer), testtype(x, DOM_IDENT);
```

```
   FALSE, TRUE
```

Queries of the following form are also possible:

```
>> assume(y > 5): is(y + 1 > 4);
```

   TRUE

Exercise 9.4: Use MuPAD to show:

$$\lim_{x \to \infty} x^a = \begin{cases} \infty & \text{for } a > 0, \\ 1 & \text{for } a = 0, \\ 0 & \text{for } a < 0. \end{cases}$$

Hint: Use the function **assume** to distinguish the cases.

10. Chance and Probability

You can use the random number generator **random** to perform many experiments in MuPAD. The call **random()** generates a random non-negative 12–digit integer. You obtain a sequence of 4 such random numbers as follows:

```
>> random(), random(), random(), random();
```

427419669081, 321110693270, 343633073697, 474256143563

If you want to generate random integers in a different range, then you can construct a random number generator **generator:=random(m..n)**. You call this generator without arguments[1], and it returns integers between m and n. The call **random(n)** is equivalent to **random(0..n-1)**. Thus you can simulate 15 rolls of a die as follows:

```
>> die := random(1..6):
>> dieExperiment := [die() $ i = 1..15];
```

[5, 3, 6, 3, 2, 2, 2, 4, 4, 3, 3, 2, 1, 4, 4]

We stress that you must specify a loop variable when using the sequence generator **$**, since otherwise **die()** is called only once and a sequence of copies of this value is generated:

```
>> die() $ 15;
```

6, 6, 6, 6, 6, 6, 6, 6, 6, 6, 6, 6, 6, 6, 6

Here is a simulation of 8 coin tosses:

---

[1] In fact, you can call **generator** with arbitrary arguments, which are ignored when generating random numbers.

```
>> coin := random(2):
>> coinTosses := [coin() $ i = 1..8];

    [0, 0, 0, 1, 1, 1, 0, 0]

>> subs(coinTosses, [0 = head, 1 = tail]);

    [head, head, head, tail, tail, tail, head, head]
```

The following example generates uniformly distributed floating-point numbers from the interval $[0, 1]$:

```
>> generator := float@random(0..10^10)/10^10:
>> randomNumbers := [generator() $ i = 1..10];

    [0.5473509389, 0.6470788275, 0.1338779163,

     0.2495900314, 0.1209539225, 0.8363442949,

     0.3746086507, 0.5862664913, 0.4813656213,

     0.384244378]
```

Here we have used `random(0..n)` to generate random integers between 0 and n, divided them by n to obtain rational numbers in $[0, 1]$, and finally converted them to floating-point numbers via `float`. The resulting floating-point numbers are actually from a discrete set. For sufficiently large n (=`10^DIGITS`), this set corresponds essentially to the set of floating-point numbers that can be represented with `DIGITS` decimal digits.

The library `stats` comprises functions for statistical analysis. You obtain information by entering `info(stats)` or `?stats`. The function `stats::mean` computes the mean value $X = \dfrac{1}{n} \sum_{i=1}^{n} x_i$ of a list of numbers $[x_1, \ldots, x_n]$:

```
>> stats::mean(dieExperiment),
   stats::mean(coinTosses),
   stats::mean(randomNumbers);

   16/5, 3/8, 0.4361681072
```

The function `stats::variance` returns the variance

$$V = \left(\frac{1}{n} \sum_{i=1}^{n} x_i^2\right) - X^2 = \frac{1}{n} \sum_{i=1}^{n} (x_i - X)^2 \; :$$

```
>> stats::variance(dieExperiment),
   stats::variance(coinTosses),
   stats::variance(randomNumbers);
```

```
122/75, 15/64, 0.04735946358
```

You can compute the standard deviation $\sqrt{V}$ with `stats::stdev`:

```
>> stats::stdev(dieExperiment),
   stats::stdev(coinTosses),
   stats::stdev(randomNumbers);
```

```
  1/2     1/2      1/2
  3       122      15
-----------,  -----, 0.2176222957
    15          8
```

If you specify the option `Sample`, then the system returns $\sqrt{\frac{n}{n-1}}\, V$ instead:

```
>> stats::stdev(dieExperiment, Sample),
   stats::stdev(coinTosses, Sample),
   stats::stdev(randomNumbers, Sample);
```

```
   1/2    1/2     1/2    1/2
  35     61      14     15
-----------,  -----------, 0.2293940413
    35            28
```

The data structure `Dom::Multiset` (see `?Dom::Multiset`) provides a simple means of determining frequencies in sequences. The call `Dom::Multiset(a,b,..)` returns a set of lists. The first entry of each such list is one of the arguments, and the second entry counts the number of occurrences in the argument sequence:

```
>> Dom::Multiset(a, b, a, c, b, b, a, a, c, d, e, d);
```

    {[a, 4], [b, 3], [c, 2], [d, 2], [e, 1]}

If you simulate 1000 rolls of a die, you might obtain the following frequencies:

```
>> rolls := die() $ i = 1..1000:
>> Dom::Multiset(rolls);
```

    {[4, 190], [2, 153], [5, 177], [1, 160], [6, 158],

        [3, 162]}

In this case you would have rolled 160 times a 1, 153 times a 2, and so on.

An example from number theory is the distribution of the greatest common divisors (gcd) of random pairs of integers. We use the function zip (Section 4.6) to combine two random lists via the function igcd, which computes the gcd:

```
>> list1 := [random() $ i=1..1000]:
>> list2 := [random() $ i=1..1000]:
>> gcdlist := zip(list1, list2, igcd);
```

    [1, 3, 6, 1, 3, 5, 1, 1, 1, 4, 1, 2, 1, 4, 1, 1, ...]

Now we use Dom::Multiset to count the frequencies of the individual gcds:

```
>> frequencies := Dom::Multiset(op(gcdlist));
```

    {[4, 28], [30, 2], [377, 1], [32, 2], [33, 1], ... }

The list becomes much more readable if we sort it by the first entry of the sublists. We employ the function sort which takes a function representing a sorting order as second argument. The latter function decides which of two elements x, y shall precede the other. We refer to the corresponding help page: ?sort. In this case, x, y are lists with two entries, and we want x to appear before y if we have $x[1] < y[1]$ (i.e., we sort numerically with respect to the first entries):

```
>> sortingOrder := (x, y) -> (x[1] < y[1]):
>> sort([op(frequencies)], sortingOrder);
```

```
    [[1, 598], [2, 142], [3, 83], [4, 28], [5, 33], ... ]
```

Thus 589 out of the 1000 chosen random pairs have a gcd of 1 and
hence are coprime. This experiment yields $59,8\%$ as an approximation
to the probability that two randomly chosen integers are coprime[2].

Exercise 10.1: We roll three dice simultaneously. For each value between 3 and 18, the following table contains the expected frequencies
of dice score (i.e., the sum of the top numbers) when rolling the dice
126 times:

score

3	4	5	6	7	8	9	10	11	12	13	14	15	16	17	18
1	3	6	10	15	21	25	27	27	25	21	15	10	6	3	1

frequency

Simulate 216 rolls and compare the frequencies that you observe with
those from the table.

Exercise 10.2: The Monte-Carlo method for approximating the area
of a region $A \subset \mathbb{R}^2$ works as follows. First, we choose a (preferably
small) rectangle Q enclosing A. Then we randomly choose n points
in Q. If m of these points lie in A, then the following holds for sufficiently large n:

$$\text{area of } A \approx \frac{m}{n} \times \text{area of } Q .$$

Let $r()$ be a generator for uniformly random numbers in the interval
$[0, 1]$. We can use this to generate uniformly distributed random vectors
in the rectangle $Q = [0, a] \times [0, b]$ via $[a*r(),b*r()]$.

a) Consider the right upper quadrant of the unit circle around the
   origin. Use Monte-Carlo simulation with $Q = [0, 1] \times [0, 1]$ to approximate its area. This way you can get random approximations
   of π.

b) Let $f : x \mapsto x \sin(x) + \cos(x) \exp(x)$. Determine an approximation
   for $\int_0^1 f(x)\,dx$. For that purpose, find an upper bound M for f on
   the interval $[0, 1]$ and apply the simulation to $Q = [0, 1] \times [0, M]$.
   Compare your result to the exact integral.

---

[2] The theoretical value of this probability is $6/\pi^2 \approx 0.6079.. \cong 60.79\%$.

11. Graphics

MuPAD provides a variety of graphical tools for rendering mathematical objects in two- and three-dimensional space. The basic functions are plotfunc[1], plot2d, and plot3d. Additionally, the library plotlib provides special plotting routines. In the following, we assume that you have a MuPAD version with a graphical user interface. Depending on the version, MuPAD's graphics module either opens a separate window after you enter a graphics command, or the graphics appears in your notebook immediately after the call, like all other results returned by MuPAD.

11.1 Graphs of Functions

The function plotfunc plots graphs of functions with one or two arguments. The command

```
>> plotfunc(sin(x), x = 0..4*PI);
```

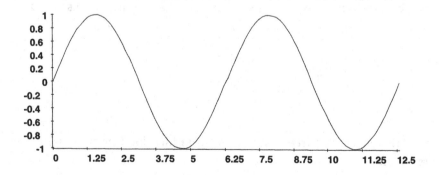

---

[1] The function plotfunc is split into plotfunc2d and plotfunc3d in MuPAD versions beyond 1.4. Their calling syntax may differ slightly from the description in this chapter; consult the help pages ?plotfunc2d and ?plotfunc3d for details.

plots the sine function in the interval $[0, 4\pi]$. You can display the graphs of several functions simultaneously in one "scene":

```
>> plotfunc(sin(x), cos(x), x = 0..4*PI);
```

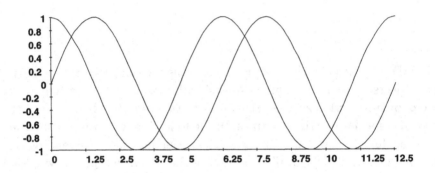

You can plot functions with singularities as well:

```
>> plotfunc(1/(1 - x) + 1/(1 + x), x = -2..2);
```

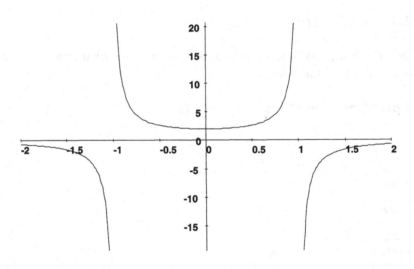

MuPAD generates three-dimensional graphics for functions with two arguments:

```
>> plotfunc(sin(x^2 + y^2), cos(x^2 + y^2),
            x = 0..PI, y = 0..PI);
```

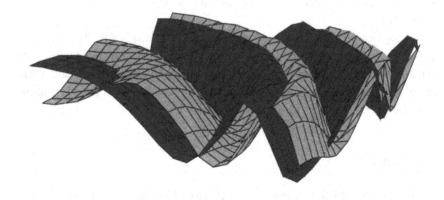

When calling plotfunc you only specify the functions and their (common) plot interval. The above three-dimensional picture shows that the default settings within plotfunc often produce unsatisfactory results: in the above case the default discretization with 20 × 20 grid points is too coarse to yield an appealing picture. You can now manipulate the graphics using the menu buttons of the graphics window: change the number of grid points, choose various kinds of axes, change colors, etc. Within a notebook, this is also possible after double-clicking on the graphics.

When you work interactively with graphics, you have a visual control of the graphics parameters. Alternatively you can directly call the more general functions plot2d or plot3d instead of plotfunc, using appropriate options. Internally plotfunc calls these functions and supplies standard values for various parameters.

11.2 Graphical Scenes

In principle a call of plot2d or plot3d looks like this:

```
>> plot2d(sceneOption1, sceneOption2, .. , Object1,
         Object2, ..);
>> plot3d(sceneOption1, sceneOption2, .. , Object1,
         Object2, ..);
```

In the following sections, we describe the structure of the various possible objects. A MuPAD sequence Object1,Object2,.. of graphical objects is called a "scene". This collection of objects is to be rendered in one graphics.

The options for a scene determine the general view: the appearance of the axes, the color of the plot, the title of the picture and its position, etc. Every scene option is a MuPAD equation of the form option=value. Table 11.1 shows all possible options. We give some examples in the following sections.

11.3 Curves

Usually the graphics objects for plot2d are *parametric curves*, i.e., you supply the x and y coordinates of the curve points as functions of a parameter. The graph of a function $f(x)$ on the interval $a \le x \le b$ can be obtained by the parameterization

$$x = u , \quad y = f(u) , \quad a \le u \le b.$$

You cannot generate a (complete) circle around the origin $(0,0)$ with radius r as the graph of a function. However, it is easy to define it in parameterized form via

$$x = r \cos(u) , \quad y = r \sin(u) , \quad 0 \le u \le 2\pi.$$

You can declare a two-dimensional parametric curve as a MuPAD list, as follows:

```
>> Object := [Mode = Curve, [x(u), y(u)], u = [a ,b],
              objectOption1, objectOption2, ..]:
```

Table 11.2 summarizes the possible options for defining a graphical object of type "curve". Like scene options, you specify them as a sequence of MuPAD equations option=value. Use MuPAD expressions for the parameterizations $x(u)$ and $y(u)$. You can generate the graph of the sine function on the interval $[0, 4\pi]$ or a circle with radius as follows:

```
>> SineGraph := [Mode = Curve, [u, sin(u)],
                 u = [0, 4*PI], Grid = [100]]:
>> Circle := [Mode = Curve, [cos(u), sin(u)],
              u = [0, 2*PI]]:
```

option	possible values	meaning	default
Arrows	TRUE/FALSE	axes as arrows?	FALSE
Axes	NONE	axes: none,	CORNER
	ORIGIN	centered at the origin,	
	CORNER	in a corner, or	
	BOX	as a frame?	
AxesOrigin	Automatic	center of the axes	Automatic
	[x0,y0]		
AxesScaling	[Lin/Log,	linear or logarithmic	[Lin,Lin]
	Lin/Log]	scale?	
BackGround	[R,G,B]	background color	[0,0,0]
	$(R,G,B \in [0,1])$		(black)
ForeGround	[R,G,B]	color for axes, title,	[1,1,1]
		etc.	(white)
FontFamily	string	font for axes labels	"Helvetica"
		and title	
FontSize	7,8,...,36	font size for axes,	10
		labels, and title	
FontStyle	string	font style for axes,	"Bold"
		labels, and title	
Labeling	TRUE/FALSE	label axes?	FALSE
Labels	[Name,Name]	axes labels	["x-axis",..]
LineStyle	SolidLines	line style	SolidLines
	DashedLines		
LineWidth	0,1,2,..	line width	0
PointStyle	Squares	point style: render	FilledSquares
	Circles	points as little	
	FilledSquares	squares, circles, etc.	
	FilledCircles		
PointWidth	1,2,3,..	point width	3
PlotDevice	Screen	output to screen,	Screen
	string	to a binary file, or	
	[string, format]	to a formatted file	
Scaling	Constrained	equal horizontal and	UnConstrained
	UnConstrained	vertical scaling?	(circles become
		(circles become either	ellipses)
		circles or ellipses)	
Ticks	0,1,...,20	(minimal) number of	10
		axes tickmarks	
Title	string	scene title	"" (no title)
TitlePosition	Above	title placement: above	Above
	Below	scene, below scene,	
	[X,Y]	at coordinates X,Y	
	$(X,Y \in [0,10])$	([0,0] = top left,	
		[10,10] = bottom right)	

Table 11.1. Options for a graphical scene

option	possible values	meaning	default
Color	[Flat] [Flat, [R,G,B]] [Height] ...	color: monochrome ("Flat") with automatically chosen or specified [R,G,B] color, color according to vertical coordinate, also user defined colors are possible	[Flat]
Grid	[2],[3],..	number of sample points	[20]
Smoothness	[0],[1],..,[20]	additional points between sample points to smooth the curve	[0]
Style	[Points] [Lines] [LinesPoints] [Impulses]	represent curve by points, a line, a line and points, or impulses	[Lines]
LineStyle	SolidLines DashedLines	line style	SolidLines
LineWidth	0,1,2,..	line width	0
PointStyle	Squares Circles FilledSquares FilledCircles	point style: render points as little squares, circles, etc.	FilledSquares
PointWidth	1,2,3,..	point width	3
Title	string	scene title	"" (no title)
TitlePosition	[X,Y] (X,Y ∈ [0,10])	title placement ([0,0] = top left, [10,10] = bottom right)	

Table 11.2. Object options for a curve

Here MuPAD computes 100 sample points for the graph of the function and interpolates linearly between them, as requested by the object option Grid=[100]. For the circle, the system uses the default value of 20 sample points.

In the following graphics commands, we specify scene options to choose the background color "white" and the foreground color (i.e., for the axes) "black", to switch off the axes ticks via Ticks=0, and to let the axes appear as a frame via Axes=Box. For all other parameters MuPAD uses the default values:

```
>> sceneOptions := BackGround = [1, 1, 1],
                   ForeGround = [0, 0, 0],
                   Ticks = 0, Axes = Box:
```

```
>> plot2d(sceneOptions, SineGraph);
```

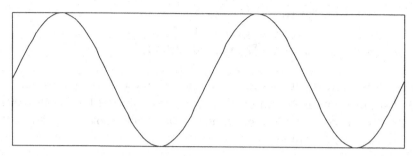

```
>> plot2d(sceneOptions, Circle);
```

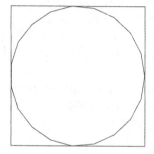

You specify colors by lists [R,G,B] of red, green, and blue values between 0.0 and 1.0. Black and white correspond to [0,0,0] and [1,1,1], respectively. The library RGB contains numerous color names with corresponding RGB values. You can list them via info(RGB):

```
>> RGB::Black, RGB::White, RGB::Red, RGB::SkyBlue;
```

```
   [0.0, 0.0, 0.0], [1.0, 1.0, 1.0], [1.0, 0.0, 0.0],

   [0.529405, 0.807794, 0.921598]
```

If you load the color library via export(RGB), then you can use the color names in the short form Black, White, etc.

The next scene comprises two objects: object 1 is a circle with radius 1 to be plotted with only few sample points. We choose monochrome (Flat) red as color scheme:

```
>> Circle1 := [Mode = Curve, [cos(u), sin(u)],
            u = [0, 2*PI], Grid = [15],
            Title = "circle 1",
            TitlePosition = [1.7, 2],
            Color = [Flat, RGB::Red]]:
```

Object 2 is a circle with radius 1/2. The parameter u runs through 100 equidistant points in the interval $[0, 2\pi]$, as requested by Grid=[100]. Due to the bigger number of samples, the circle looks very smooth. The chosen color is blue:

```
>> Circle2 := [Mode = Curve, [cos(u)/2, sin(u)/2],
            u = [0, 2*PI], Grid = [100],
            Title = "circle 2",
            TitlePosition = [6.3, 2.8],
            Color = [Flat, RGB::Blue]]:
```

In the scene options, we supply a MuPAD string for the title. We choose a larger font for the labels, axes centered at the origin, and switch on axes labels by Labeling=TRUE:

```
>> sceneOptions :=
        Title = "two circles", FontSize = 14,
        TitlePosition = [5, 0.7], Axes = Origin,
        Labeling = TRUE, Labels = ["x", "y"],
        BackGround = [1, 1, 1], ForeGround = [0, 0, 0]:
```

You can see clearly the effect of the different object options in the graphics:

```
>> plot2d(sceneOptions, Circle1, Circle2);
```

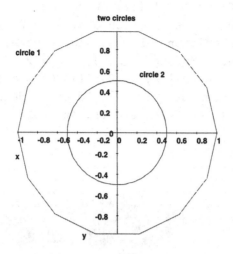

11.4 Surfaces.

The typical graphic objects in three space for **plot3d** are *parametric surfaces*: the x, y, z coordinates of the surface points are defined as functions $x(u, v)$, $y(u, v)$, $z(u, v)$ of two surface parameters u, v. In this way, you can define the graph of a function $f(x, y)$ in two variables on the rectangle $a \leq x \leq b$, $A \leq y \leq B$ in the form

$$x = u , \quad y = v , \quad z = f(u, v) , \quad a \leq u \leq b , \quad A \leq v \leq B.$$

You can easily plot more complex surfaces such as the surface of a sphere with radius r in the parametric form

$$x = r \cos(u) \sin(v) , \quad y = r \sin(u) \sin(v) , \quad z = r \cos(v) ,$$
$$0 \leq u \leq 2\pi , \ 0 \leq v \leq \pi .$$

You define a parametric surface as a MuPAD list as follows:

```
>> Object := [Mode = Surface,
              [x(u, v), y(u, v), z(u, v)],
              u = [a, b], v = [A, B],
              objectOption1, objectOption2, ..
             ]:
```

For example, the following command plots the graph of the function $f(x, y) = \sin(x^2 + y^2)$:

```
>> plot3d(BackGround = [1,1,1], ForeGround = [0,0,0],
        Title = "the function sin(x^2+y^2)",
        TitlePosition = Below, Axes = None,
        [ Mode = Surface, [x, y, sin(x^2 + y^2)],
          x = [0, PI], y = [0, PI], Grid = [20, 20],
          Smoothness = [2, 2], Color = [Height],
          Style = [ColorPatches, AndMesh]
        ]);
```

the function sin(x^2+y^2)

In this example, we have chosen 20 equidistant sample points for each of the two surface parameters u and v and two further intermediate points between two neighboring sample points for smoothing (Smoothness=[2,2]). Thus the system has computed $60 \times 60 = 3600$ surface points in total. The title appears under the picture (TitlePosition=Below). The color of each surface point is determined by its z value (Color=[Height]).

The object options for surfaces are similar to those for curves (Table 11.2). However, the values of Grid and Smoothness are now lists with two entries, so that you can specify the number of sample points and the number of intermediate points for smoothing independently for both surface parameters. The option Style can assume the values given in Table 11.3 for surfaces. The default is Style=[Wireframe,Mesh]. The scene options for plot3d agree with those for plot2d[2] (Table 11.1), plus some additional options:

---

[2] Specify the parameters for the axes by lists with three elements, for example: Labels=["x","y","z"], AxesOrigin=[0,0,0], AxesScaling=[Lin,Lin,Lin].

Style =	surface representation
[Points]	only points
[WireFrame,Mesh]	wire frame of both parameter lines
[WireFrame,ULine]	wire frame of first parameter lines
[WireFrame,VLine]	wire frame of second parameter lines
[HiddenLine,Mesh]	opaque surface with both parameter lines
[HiddenLine,ULine]	opaque surface with first parameter lines
[HiddenLine,VLine]	opaque surface with second parameter lines
[ColorPatches,Only]	colored surface without parameter lines
[ColorPatches,AndMesh]	colored surface with both parameter lines
[ColorPatches,AndULine]	colored surface with first parameter lines
[ColorPatches,AndVLine]	colored surface with second parameter lines
[Transparent,Only]	transparent surface without parameter lines
[Transparent,AndMesh]	transparent surface with both parameter lines
[Transparent,AndULine]	transparent surface with first parameter lines
[Transparent,AndVLine]	transparent surface with second parameter lines

Table 11.3. Style options for surfaces

3D scene option	possible values	meaning	default
CameraPoint	Automatic [x,y,z]	position of the observer	Automatic
FocalPoint	Automatic [x,y,z]	focal point of the observer	Automatic

This determines the parameters for the perspective of a three-dimensional image: a camera resides at the point CameraPoint, and its optical axis points in the direction of the point FocalPoint in three space.

11.5 Further Possibilities

Up to now we have presented graphical objects of type "curve" (in two dimensions) and of type "surface" (in three dimensions). In addition plot3d provides the possibility to plot space curves (Mode=Curve). Moreover, both plot2d and plot3d can render lists of graphical primitives (points or polygons). You can use this to visualize discrete data. Pass the points as a list [point(x1,y1),point(x2,y2),..] to plot2d or as a list [point(x1,y1,z1),point(x2,y2,z2),..] to plot3d:

```
>> PlotPoints := [point(i, sin(i*6.28/50)) $ i = 0..50]:
>> plot2d(BackGround = [1, 1, 1], ForeGround = [0, 0, 0],
          Scaling = UnConstrained, PointWidth = 30,
          [Mode = List, PlotPoints,
           Color = [Flat, [0, 0, 0]]]);
```

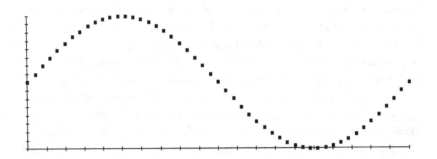

The library plotlib[3] provides advanced algorithms for graphical representations:

```
>> info(plotlib);
```

```
    Library 'plotlib': Library routines for two- and
    three-dimensional plots.
    Interface:
    plotlib::contourplot,    plotlib::cylindricalplot,
    plotlib::dataplot,       plotlib::densityplot,
    plotlib::fieldplot,      plotlib::implicitplot,
    plotlib::polarplot,      plotlib::polygonplot,
    plotlib::sphericalplot, plotlib::xrotate,
    plotlib::yrotate
```

We refer to the corresponding help pages for a detailed description and give some examples below.

The function plotlib::contourplot[4] plots contour lines of a surface in three-dimensional space (see ?plotlib::contourplot). In the

---

[3] The plotlib library is redesigned for MuPAD versions beyond 1.4. Please consult the help page ?plotlib.

[4] The function plotlib::contourplot is replaced by plotlib::implicitplot in MuPAD versions beyond 1.4. Consult the help page ?plotlib::implicitplot for further information.

following example, we illustrate the solution curve of all points (x, y)
satisfying the equation

$$f(x, y) = (x^2 + y^2)^3 - (x^2 - y^2)^2 = const., \quad -1 \leq x \leq 1, \; -1 \leq y \leq 1.$$

We use `plotlib::contourplot` to plot the contour lines of the function $f(x, y)$. The following example MuPAD simultaneously plots the
solution curves for $f(x, y) = -0.05$, $f(x, y) = 0$, and $f(x, y) = 0.05$.
The "cloverleaf" curve corresponds to the solutions of $f(x, y) = 0$:

```
>> sceneOptions :=  Axes = Box, Labeling = TRUE,
                    Labels = ["",""],
                    BackGround = [1, 1, 1],
                    ForeGround = [0, 0, 0]:
>> mySurface := [x, y, (x^2 + y^2)^3 - (x^2 - y^2)^2],
                x = [-1, 1], y = [-1, 1]:
>> Options :=  Contours = [-0.05, 0, 0.05],
               Grid = [100, 100]:
>> plotlib::contourplot(sceneOptions,
                        [mySurface, Options]);
```

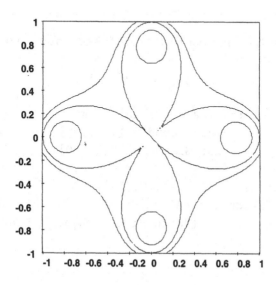

The function `plotlib::xrotate` generates the rotational surface that
results from rotating a curve around the x axis:

```
>> myCurve := [x, sin(x)], x = [0, 3*PI]:
>> objectOptions := Color = [Height], Grid = [100, 20],
                    Style = [ColorPatches, AndMesh]:
>> RotationalSurface := [myCurve, angle = [0, 2*PI],
                         objectOptions]:
>> sceneOptions :=  Axes = Box, BackGround = [1, 1, 1],
                    ForeGround = [0, 0, 0]:
>> plotlib::xrotate(sceneOptions, RotationalSurface);
```

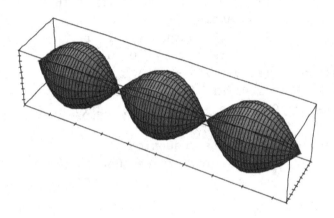

Finally we use the function plotlib::fieldplot to plot the vector field

$$(x, y) \;\rightarrow\; \left(-y^2, x^2\right):$$

```
>> sceneOptions := Axes = Origin, Ticks = 0,
                   BackGround = [1, 1, 1],
                   ForeGround = [0, 0, 0]:
>> VectorField := [-y^2, x^2], x = [-4, 4], y = [-4, 4]:
>> objectOptions := Grid = [20, 20]:
>> plotlib::fieldplot(sceneOptions,
                      [VectorField, objectOptions]);
```

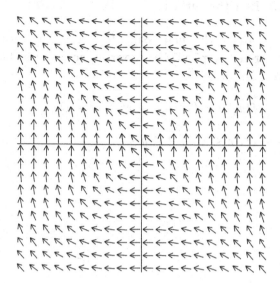

You find a more detailed description of MuPAD's graphics features and further examples in the MuPAD User's Manual [MuP 96], in the demonstration [DPS 97], or on the help pages for `plot2d` and `plot3d`.

11.6 Printing and Saving Graphics

You can print and save graphics interactively. Click on the corresponding button of the graphics window. Then you can choose various graphics formats such as Postscript or GIF from a menu. The system generates an external file containing the graphical data in the selected format. The data in this file can then be used by other software or be sent to an appropriate printer. Alternatively, you can use the scene option `PlotDevice` (see Table 11.1) within a plotting command to generate an external file.

When you print color graphics on a black and white printer, colors are often simulated by grayscale values. This may lead to unsatisfactory results. In such cases, we recommend to let MuPAD generate black and white graphics directly.

Exercise 11.1: Let rd denote the rounding function mapping a real point x to the nearest integer. Plot the function $f(x) = \dfrac{|x - rd(x)|}{x}$ on the inverval $[1, 30]$.

Exercise 11.2: Plot the surface of a sphere in MuPAD.

Exercise 11.3: Read the help page for `plotlib::implicitplot`. What is the difference to `plotlib::contourpolot`? Use both routines to plot the solution set of

$$f(x, y) = (x^2 + y^2)(x^2 + y^2 - 1) = 0$$

on the square $-2 \leq x \leq 2$, $-2 \leq y \leq 2$.

12. The History Mechanism

Every input to MuPAD yields a result after evaluation by the system. The computed objects are stored internally in a *history table*. Note that the result of any statement is stored, even if it is not printed on the screen. You can it them later by means of the function `last`. The command `last(1)` returns the previous result, `last(2)` the last but one, and so on. Instead of `last(i)` you may use the shorter notation `%i`. Moreover, `%` is short for `%1` or `last(1)`. Thus the input

```
>> f := diff(ln(ln(x)), x): int(f, x);
```

can be passed to the system in the following equivalent form:

```
>> diff(ln(ln(x)), x): int(%, x);
```

This enables you to access intermediate results that have not been assigned to an identifier. It is remarkable that the use of `last` may speed up certain evaluations. In the following example, we first try to compute a definite integral symbolically. After recognizing that Mu-PAD does not compute a symbolic value, we ask for a floating-point approximation:

```
>> f := int(sin(x)*exp(x^3)+x^2*cos(exp(x)), x=0..1);
```

$$\text{int(sin(x) exp(x}^3\text{) - 2 x}^2 \text{ cos(exp(x))), x = 0..1)}$$

```
>> startingTime := time():
>> float(f);
```

```
1.051120738
```

```
>> time() - startingTime;
```

    1150

The function `time` returns the total computing time (in milliseconds)
used by the system since the beginning of the session. Thus the printed
difference is the time for computing the floating-point approximation.
In this example, we can reduce the computing time dramatically by
employing `last`:

```
>> f := int(sin(x)*exp(x^3)+x^2*cos(exp(x)), x=0..1);
```

$$\mathrm{int}(\sin(x)\ \exp(x^3) - 2\ x^2\ \cos(\exp(x)),\ x = 0..1)$$

```
>> startingTime := time():
>> float(%2);
```

    1.051120738

```
>> time() - startingTime;
```

    270

In this case the reason for the gain in speed is that MuPAD does *not*
re-evaluate the objects that `last(i)`, `%i`, or `%` refer to[1]. Thus calls to
`last` form an exception to the usual complete evaluation at interactive
level (Section 5.2):

```
>> unassign(x): sin(x): x := 0: %2;
```

    sin(x)

You can enforce complete evaluation by using `eval`:

```
>> unassign(x): sin(x): x := 0: eval(%2);
```

    0

---

[1] Note that this gain in speed is only achieved when working interactively, since
identifiers are evaluated with level 1 within procedures (Section 18.10).

Please note that the value of last(i) may differ from the *i*th but last *visible* output, if you have suppressed the screen output of some intermediate results by terminating the corresponding commands with a colon. Also note that the value of the expression last(i) changes permanently during a computation:

```
>> 1: last(1) + 1; last(1) + 1;
```

```
2
```

```
3
```

The environment variable HISTORY determines the number of results that MuPAD stores in a session and that can be accessed via last:

```
>> HISTORY;
```

```
[20, 3]
```

This default means that MuPAD stores the previous 20 expressions interactively, while you can use only last(1), last(2), and last(3) within procedures (Chapter 18). Of course you can change this default by assigning a list with other values to HISTORY. This may be appropriate when MuPAD has to handle huge objects (such as very large matrices) that fill up a significant part of the main memory of your computer. Copies of these objects are stored in the history table, requiring additional storage space. In this case, you would reduce the memory load by choosing small values in HISTORY. In MuPAD versions beyond 1.4, HISTORY yields only the value of the interactive "history depth" and the user cannot change the corresponding value 3 for procedures.

We strongly recommend to use last only interactively. The use of last within procedures is considered bad programming style and should be avoided.

13. Input and Output

13.1 Output of Expressions

MuPAD does not display all computed results on the screen. Typical examples are the commands within for loops (Chapter 16) or procedures (Chapter 18): only the final result (i.e., the result of the last command) is printed and the output of intermediate results is suppressed. Nevertheless you can let MuPAD print intermediate results or change the output format.

13.1.1 Printing Expressions on the Screen

The function print outputs MuPAD objects on the screen:

```
>> for i from 4 to 5 do
     print("The ", i, "th prime is ", ithprime(i))
   end_for;

   "The ", 4, "th prime is ", 7

   "The ", 5, "th prime is ", 11
```

We recall that ithprime(i) computes the ith prime. MuPAD encloses the text in double quotes. Use the option Unquoted to suppress this:

```
>> for i from 4 to 5 do
     print(Unquoted,
           "The ", i, "th prime is ", ithprime(i))
   end_for;

   The , 4, th prime is , 7

   The , 5, th prime is , 11
```

Furthermore, you can eliminate the commas from the output by means of the tools for manipulating strings that are presented in Section 4.11:

```
>> for i from 4 to 5 do
     print(Unquoted,
            "The " . expr2text(i) . "th prime is " .
            expr2text(ithprime(i)) . ".")
   end_for;

   The 4th prime is 7.

   The 5th prime is 11.
```

Here the function `expr2text` converts the values of i and `ithprime(i)` to strings. Then the concatenation operator . combines them with the other strings to a single string.

Alternatively you can use the function **fprint**, which writes data to a file or on the screen. In contrast to **print**, it does not output its arguments as individual expressions. Instead, **fprint** combines them to a single string (if you use the option Unquoted):

```
>> a := one: b := string:
>> fprint(Unquoted, 0, "This is ", a, " ", b);

   This is one string
```

The second argument 0 tells **fprint** to direct its output to the screen.

13.1.2 Modifying the Output Format

Usually MuPAD prints expressions in a two-dimensional form with simple (ASCII) characters:

```
>> diff(sin(x)/cos(x), x);

        2
     sin(x)
     ------- + 1
        2
     cos(x)
```

This format is known as *pretty print*. It resembles the usual mathematical notation. Therefore it is often easier to read than a single line output. However, MuPAD only uses pretty print for output and it is not a valid input format: in a graphical user interface you cannot copy some output text with the mouse and paste it as MuPAD input somewhere else.

The environment variable PRETTY_PRINT[1] controls the output format. The default value of the variable is TRUE, i.e., the pretty print format is used for the output. If you set this variable to FALSE, then you obtain a one-dimensional output form which can also be used as input:

```
>> PRETTY_PRINT := FALSE: diff(sin(x)/cos(x), x);

   cos(x)^(-2)*sin(x)^2 + 1
```

If an output would exceed the line width, then the system automatically breaks the lines:

```
>> PRETTY_PRINT := TRUE: taylor(sin(x), x = 0, 17);

        3    5     7       9        11          13
       x    x     x       x        x           x
   x - -- + --- - ---- + ------ - -------- + ----------- -
       6    120   5040   362880   39916800   6227020800


        15
       x                  17
   ------------- + O(x   )
   1307674368000
```

You can set the environment variable TEXTWIDTH to the desired line width. Its default value is 75 (characters), and you can assign any integer between 9 and $2^{31} - 1$ to it. For example, if you compute $(\ln \ln x)''$, then you obtain the following output:

```
>> diff(ln(ln(x)), x, x);
```

---

[1] This variable is renamed to PRETTYPRINT in MuPAD versions beyond 1.4.

```
        1                 1
   -  --------  -  ----------

       2              2       2
      x   ln(x)      x   ln(x)
```

If you reduce the value of TEXTWIDTH, then the system breaks the output across two lines:

```
>> TEXTWIDTH := 20: diff(ln(ln(x)),x,x);
```

```
         1
   -  --------  -

        2
       x   ln(x)
```

```
         1
      ----------

       2       2
      x   ln(x)
```

The default value is restored by unassigning TEXTWIDTH:

```
>> unassign(TEXTWIDTH):
```

You can also control the output by user-defined preferences. This is discussed in Chapter 14.1.

13.2 Reading and Writing Files

You can save the values of identifiers or a complete MuPAD session to a file and read the file later into another MuPAD session.

13.2.1 The Functions write and read

The function write stores the values of identifiers in a file, so that you can reuse the computed results in another MuPAD session. In the following example we save the values of the identifiers a and b to the file ab.mb:

```
>> a := 2/3: b := diff(sin(cos(x)), x):
>> write("ab.mb", a, b);
```

You pass the file name as a string (Section 4.11) enclosed in double quotes ". The system then creates a file with this name (without "). If you read this file into another MuPAD session via the function read, then you can access the values of the identifiers a and b without re-computing them:

```
>> reset(): read("ab.mb"): a, b;

    2/3, - sin(x) cos(cos(x))
```

If you use the function write as in the above example, then it creates a file in the MuPAD specific binary format. By convention, a file in this format should have the file name extension ".mb". You can call the function write with the option Text. This generates a file in a readable text format[2]:

```
>> a := 2/3: b := diff(sin(cos(x)), x):
>> write(Text, "ab.mu", a, b);
```

The file ab.mu now contains the following two syntactically correct MuPAD commands:

```
sysassign(a, hold(2/3)):
sysassign(b, hold(-sin(x)*cos(cos(x)))):
```

You can use the function read to read this file:

```
>> a := 1: b := 2: read("ab.mu"): a, b;

    2/3, -sin(x) cos(cos(x))
```

The function sysassign works like _assign and the assignment opera-tor :=. It assigns a value to an identifier. Thus the identifiers a,b as-sume the values that they had at the execution time of the write command.

The text format files generated by write contain valid MuPAD commands. Of course, you can use any editor to generate such a text file "by hand" and read it into a MuPAD session. In fact this is a natural tool when you develop more complex MuPAD procedures.

---

[2] Usually the file name extension for MuPAD text format files should be ".mu".

13.2.2 Saving a MuPAD Session

If you call the function write without supplying any identifiers as arguments, then the system writes the values of *all* identifiers having a value to a file. Thus it is possible to restore the state of the current session via read at a later time:

```
>> result1 := .. ; result2 := .. ; ...
>> write("results.mb");
```

The function protocol records the inputs and the screen outputs of a session in a file. The command protocol(file) creates a text format file. Inputs and outputs are written to this file until you enter protocol():

```
>> protocol("logfile"):
>> limit(sin(x)/x, x = 0);
```

```
   1
```

```
>> protocol():
```

These commands generate a text file named logfile with the following contents:

```
        >> limit(sin(x)/x, x = 0);
```

```
                        1
```

```
        >> protocol():
```

It is not possible to read a file generated by protocol into a MuPAD session.

13.2.3 Reading Data from a Text File

Often you want to use data in MuPAD that are generated by a different software (for example, you might want to read in statistical values for further processing) or access all files in some directory automatically. This is possible with the help of the function io::readdata from the library io for input and output[3]. This function converts the contents

---

[3] In MuPAD versions beyond 1.4 you find this function in the library import.

of a file to a nested MuPAD list. You may regard the file as a "matrix" with line breaks indicating the beginning of a new row. Note that the rows may have different length. You may pass an arbitrary character to io::readdata as a column separator. For example, assume that the file named **directory** contains the following three lines:

```
prog1.mu
myLetter.txt
abc
```

If you take the period as column separator, you obtain:

```
>> io::readdata("directory", ".");
```

```
[[prog1, mu], [myLetter, txt], [abc]]
```

Suppose you have a file **numericalData** with the following 4 lines:

```
1    1.2    12
2.34   234
    34   345.6
4     44        444
```

By default, blank characters are assumed as column separators. So you can read this file into a MuPAD session as follows:

```
>> data := io::readdata("numericalData"):
>> data[1]; data[2]; data[3]; data[4];
```

```
[1, 1.2, 12]
```

```
[2.34, 234]
```

```
[34, 345.6]
```

```
[4, 44, 444]
```

The help page for io::readdata provides further information.

14. Utilities

In this chapter we present some useful functions. Due to space limitations, we do not explain their complete functionality and refer to the help pages for more detailed information.

14.1 User-Defined Preferences

You can customize MuPAD's behavior by using *preferences*. The following command lists all preferences:

```
>> Pref();
```

```
    Pref::noProcRemTab    : FALSE
    Pref::report          : 0
    Pref::floatFormat     : "f"
    Pref::echo            : TRUE
    Pref::prompt          : TRUE
    Pref::postOutput      : NIL
    Pref::debugOnTheFly   : TRUE
    Pref::typeCheck       : Interactive
    Pref::printTimesDot   : FALSE
    Pref::keepOrder       : DomainsOnly
    Pref::verboseRead     : 0
    Pref::maxTime         : 0
    Pref::moduleTrace     : FALSE
    Pref::maxMem          : 0
    Pref::postInput       : NIL
    Pref::promptString    : ">> "
    Pref::trailingZeroes  : FALSE
    Pref::matrixSeparator : ", "
    Pref::kernel          : [1, 4, 1]
```

```
Pref::userOptions    : ""
Pref::output         : NIL
Pref::callOnExit     : NIL
Pref::callBack       : NIL
```

We refer to the help page ?**Pref** for a complete description of all preferences and discuss some specific options below.

You can use the **report** preference to request regularly information on MuPAD's allocated memory, the memory really used, and the elapsed computing time. Valid arguments for **report** are integers between 0 and 9. The default value 0 means that no information is displayed. If you choose the value 9, then you permanently obtain information about MuPAD's current state.

```
>> Pref::report(9): limit(sin(x)/x, x = 0);

    [used=1394k, reserved=1535k, seconds=0]
                    . . .
    [used=2121k, reserved=2277k, seconds=4]
```

                    1

The preference **floatFormat** controls the output of floating-point numbers. For example, if you supply the argument "e" then floating-point numbers are printed with mantissa and exponent (e.g., $1.234\text{e-}7$ $= 1.234 \cdot 10^{-7}$). If you use the argument "f" then MuPAD prints them in fixed point representation:

```
>> Pref::floatFormat("e"): float(exp(-50));

    1.928749847e-22

>> Pref::floatFormat("f"): float(exp(-50));

    0.00000000000000000000001928749847
```

More generally, you can use preferences to control the screen output. For example, after calling **Pref::output(F)** MuPAD passes every result computed by the kernel to the function F before printing it on the screen. The screen output is then the result of the function F instead of the result originally computed by the kernel. In the following example, we use this to compute and output the normalization of every requested expression. We define a procedure F (Chapter 18) and pass it to Pref::output:

```
>> F := proc(x) begin (x, normal(x)); end_proc:
>> Pref::output(F): 1+x/(x+1)-2/x;
```

```
                  2
    x      2    2 x  - x - 2
  ----- - - + 1, -------------
  x + 1   x           2
                   x + x
```

The library **generate** contains functions for converting MuPAD expressions to the input format of other programming languages (C, Fortran, or TEX). In the following example the MuPAD output is converted to a string. You might then write this string to a text file for further processing with TEX:

```
>> Pref::output(generate::TeX): diff(f(x),x);
```

```
    "\\frac{d}{dx} f\\left(x\\right)"
```

The following command resets the output routine to its original state:

```
>> Pref::output(NIL):
```

Some users want to obtain information on certain characteristics of all computations, such as computing times. This can be achieved with the functions Pref::postInput and Pref::postOutput. Both take MuPAD procedures as arguments, which are then called after each input or output, respectively. In the following example, we use a procedure that assigns the system time returned by time() to the global identifier Time. This starts a "timer" before each computation:

```
>> Pref::postInput(proc() begin Time := time() end_proc):
```

We define a procedure **myInformation** which—among other things— uses this timer to determine the time taken by the computation. It employs **expr2text** (Section 4.11) to convert the numerical time value to a string and concatenates it with some other strings. Moreover, the procedure uses **domtype** to find the domain type of the object and converts it to a string as well. Finally it concatenates the time information, some blanks " ", and the type information via _concat. The function **strlen**, which is renamed to **length** in MuPAD versions beyond 1.4, is used to determine the precise number of blanks so that the domain type appears flushed right on the screen:

```
>> myInformation := proc() begin
      "domain type : ". expr2text(domtype(args()));
      "time : ". expr2text(time() - Time). " msec";
      _concat(%1,
               " " $ TEXTWIDTH-1-strlen(%1)-strlen(%2),
               %2)
   end_proc:
```

We pass this procedure as argument to **Pref::postOutput**:

```
>> Pref::postOutput(myInformation):
```

After each computed result, the system now prints the string generated by the procedure **myInformation** on the screen:[1]

```
>> Factor(x^3 - 1);
```

$$(x - 1)\ (x + x^2 + 1)$$

```
time : 20 msec                     domain type : DOM_EXPR
```

You can reset a preference to its default value by specifying NIL as argument. For example, the command **Pref::report(NIL)** resets the value of **Pref::report** to 0. Similarly, **Pref(NIL)** resets *all* preferences to their default values.

In MuPAD versions prior to 1.4 you have to write **pref** instead of **Pref**. Moreover, the syntax is somewhat different: consult the corrsponding help page via **?pref**.

---

[1] The function **Factor** is replaced by **factor** in MuPAD versions beyond 1.4; see the footnote on page 135.

Exercise 14.1: The MuPAD function **bytes** returns the amount of logical and physical memory used by the current MuPAD session. Use Pref::postOutput to let this information appear on the screen after each output.

14.2 Information on **MuPAD** Algorithms

Some of MuPAD's system functions may provide additional runtime information. The following command makes all procedures produce such information on the screen:

```
>> setuserinfo(Any, 1):
```

As an example, we invert the following matrices (Section 4.15) over the ring of integers modulo 11:

```
>> M := Dom::Matrix(Dom::IntegerMod(11)):
>> A := M([[1,2,3], [3,2,1], [4,5,7]]):
>> B := M([[4,5,7], [3,2,1], [1,2,3]]):
>> C := M([[3,2,1], [4,5,7], [1,2,3]]):
>> A^(-1);

   perform (ordinary) Gaussian elimination

   +-                                  -+
   |  6 mod 11, 8 mod 11, 1 mod 11  |
   |                                   |
   |  7 mod 11, 4 mod 11, 9 mod 11  |
   |                                   |
   |  1 mod 11, 2 mod 11, 1 mod 11  |
   +-                                  -+
```

You obtain more detailed information by increasing the second argument of **setuserinfo** (the "information level")[2]:

```
>> setuserinfo(Any, 3): B^(-1);
```

---

[2] If you call A^(-1) for the second time, then the system returns only the result, without any additional information. The reason is that the inversion procedure is implemented with the **option remember**, such that the result is taken directly from the remember table without running the algorithm again.

```
perform (ordinary) Gaussian elimination
search for pivot
  in column , 1
pivot element is , 4 mod 11
search for pivot
  in column , 2
pivot element is , 1 mod 11
```

```
+-                              -+
|  1 mod 11, 8 mod 11, 6 mod 11  |
|                                |
|  9 mod 11, 4 mod 11, 7 mod 11  |
|                                |
|  1 mod 11, 2 mod 11, 1 mod 11  |
+-                              -+
```

If you enter

```
>> setuserinfo(Any, 0):
```

then the system stops printing additional information:

```
>> C^(-1);
```

```
+-                              -+
|  8 mod 11, 1 mod 11, 6 mod 11  |
|                                |
|  4 mod 11, 9 mod 11, 7 mod 11  |
|                                |
|  2 mod 11, .1 mod 11, 1 mod 11  |
+-                              -+
```

The first argument of **setuserinfo** may be an arbitrary procedure name or library name. Then the corresponding procedure(s) provide additional information. Programmers of the system functions have built output commands in the code via **userinfo**. These commands are activated by **setuserinfo**. You can use this in your own procedures as well (**?userinfo**).

14.3 Restarting a **MuPAD** Session

The command `reset()` resets a MuPAD session to its initial state. Afterwards all identifiers that you defined previously have no value and all environment variables are reset to their default values:

```
>> a := hello: DIGITS := 100: reset(): a, DIGITS;

   a, 10
```

14.4 Executing Commands of the Operating System

You can use the function **system**, or the exclamation symbol ! for short, to execute a command of the operating system. On UNIX platforms the following command lists the contents of the current directory:

```
>> !ls

   !ls /user/dummy/MuPAD/share
   changes/    demo/   examples/   mmg/    xview/   bin/
   copyright/  doc/    lib/        tex/
```

You can neither use the output of such a command for further computation nor save it to a file[3]. The result returned by **system** is the error status that the operating system reports for the command.

The function **system** is not available on all platforms. For example, you can neither use it on a Windows system nor on a Macintosh.

---

[3] If this is desired, then you can use another command of the operating system to write the output to a file and read this file into a MuPAD session via `io::readdata`.

15. Type Specifiers

The data structure of a MuPAD object is its domain type, which can be requested by means of the function domtype. The domain type reflects the structure that the MuPAD kernel uses internally to manage the objects. The type concept also leads to a classification of the objects according to their mathematical meaning: numbers, sets, expressions, series expansions, polynomials, etc.

In this section we describe how to obtain detailed information about the mathematical structure of objects. For example, how can you find out efficiently whether an integer of domain type DOM_INT is *positive* or *even*, or whether all elements of a set are equations?

Such type checks are barely relevant for interactive use of MuPAD, since then you can control the mathematical meaning of an object by direct inspection. Type checks are mainly used for implementing mathematical algorithms, i.e., when programming MuPAD procedures (Chapter 18). For example, a procedure for differentiating expressions has to decide whether its input is a product, a composition of functions, a symbolic call of a known function, etc. Each case requires a different action, such as spplying the product rule, the chain rule, etc.

15.1 The Functions type and testtype

For most MuPAD objects the function type returns, like domtype, the domain type:

```
>> type([a, b]), type({a, b}), type(array(1..1));
```

```
    DOM_LIST, DOM_SET, DOM_ARRAY
```

For expressions of domain type DOM_EXPR the function type yields a finer distinction according to the mathematical meaning of the expression: sums, products, function calls, etc.:

```
>> type(a + b), type(a*b), type(a^b), type(a(b));
```

   "_plus", "_mult", "_power", "function"

```
>> type(a = b), type(a < b), type(a <= b);
```

   "_equal", "_less", "_leequal"

The result returned by **type** is the function call that generates the expression (internally a symbolic sum or product is represented by a call of the function _plus or _mult, respectively). More generally the result for a symbolic call of a system function is the identifier of the function as a string:

```
>> type(ln(x)), type(diff(f(x), x)), type(fact(x));
```

   "ln", "diff", "fact"

You can use both the domain types DOM_INT, DOM_EXPR, etc. and the strings returned by **type** as *type specifiers*. There exists a variety of other type specifiers in addition to the "standard typing" of MuPAD objects given by **type**. An example is NUMERIC[1]. This type comprises all "numerical" objects (of domain type DOM_INT, DOM_RAT, DOM_FLOAT, or DOM_COMPLEX).

   The call **testtype(object,typeSpecifier)** checks whether an object complies with the specified type. The result is either **TRUE** or **FALSE**. Several type specifiers may correspond to an object:

```
>> testtype(2/3, DOM_RAT), testtype(2/3, NUMERIC);
```

   TRUE, TRUE

```
>> testtype(2 + x, "_plus"), testtype(2 + x, DOM_EXPR);
```

   TRUE, TRUE

```
>> testtype(f(x), "function"), testtype(f(x), DOM_EXPR);
```

   TRUE, TRUE

---

[1] In MuPAD versions beyond 1.4, NUMERIC is replaced by Type::Numeric.

Exercise 15.1: Consider the expression

$$f(i) \;=\; \frac{i^{5/2} + i^2 - i^{1/2} - 1}{i^{5/2} + i^2 + 2\,i^{3/2} + 2\,i + i^{1/2} + 1}$$

How can MuPAD decide whether the set

```
>> S := {f(i) $ i = -1000..-2} union {f(i) $ i=0..1000}:
```

contains only rational numbers? Hint: For a specific integer i use the function normal to simplify subexpressions of f(i) containing square roots.

Exercise 15.2: Consider the expressions $\sin(i\,\pi/200)$ for integral values of i between 0 and 100. Which of them are simplified by MuPAD's sin function, which are returned as symbolic values sin(..)?

15.2 Comfortable Type Checking: the Type Library

The type specifiers presented above are useful only for checking relatively simple structures. For example, how can you check without direct inspection whether the object [1,2,3,..] is a list of positive integers?

For that purpose the Type library provides further type specifiers and constructors. You can use them to create your own type specifiers, which are recognized by testtype:

```
>> info(Type);
```

```
Domain 'Type': Type expressions for testing types
Interface:
Type::AnyType,        Type::Complex,
Type::Divs,           Type::Even,
Type::Fraction,       Type::IV,
Type::Imaginary,      Type::IntImaginary,
Type::Integer,        Type::Irrational,
Type::ListOf,         Type::ListOfIdents,
Type::ListProduct,    Type::NegInt,
Type::NegRat,         Type::Negative,
Type::NonNegInt,      Type::NonNegRat,
```

```
Type::NonNegative,  Type::Odd,
Type::PolyOf,       Type::PosInt,
Type::PosRat,       Type::Positive,
Type::Prime,        Type::Product,
Type::Rational,     Type::RealNum,
Type::Relation,     Type::SequenceOf,
Type::Series,       Type::SetOf,
Type::Singleton,    Type::TableOfEntry,
Type::TableOfIndex, Type::Union,
Type::Zero
```

For example, the type specifier `Type::PosInt` represents the set of positive integers $n > 0$, `Type::NonNegInt` corresponds to the nonnegative integers $n \geq 0$, `Type::Even` and `Type::Odd` represent the even and odd integers, respectively. These type specifiers are of domain type `Type`:

```
>> domtype(Type::Even);
```

```
   Type
```

You can use such type specifiers to query the mathematical structure of MuPAD objects via `testtype`. In the following example, we extract all even integers from a list of integers via `select` (Section 4.6):

```
>> select([i $ i = 1..20], testtype, Type::Even);
```

```
   [2, 4, 6, 8, 10, 12, 14, 16, 18, 20]
```

You can use constructors such as `Type::ListOf` or `Type::SetOf` to perform type checking for lists or sets: a list of integers is of type `Type::ListOf(DOM_INT)`, a set of equations corresponds to the type specifier `Type::SetOf("_equal")`, and a set of odd integers is of type `Type::SetOf(Type::Odd)`.

```
>> T := Type::ListOf(DOM_INT):
>> testtype([-1, 1], T), testtype({-1, 1}, T),
   testtype([-1, 1.0], T);
```

```
   TRUE, FALSE, FALSE
```

The constructor `Type::Union` generates type specifiers corresponding to the union of simpler types. For example, the following type specifier

```
>> T := Type::Union(DOM_FLOAT, Type::NegInt, Type::Even):
```

represents the union of the set of floating-point numbers, the set of negative integers, and the set of even integers:

```
>> testtype(-0.123, T), testtype(-3, T),
   testtype(2, T), testtype(3, T);

   TRUE, TRUE, TRUE, FALSE
```

We describe an application of type checking for the implementation of MuPAD procedures in Section 18.6.

Exercise 15.3: How can you compute the intersection of a set with the set of positive integers?

Exercise 15.4: Use ?`Type::ListOf` to consult the help page for this type constructor. Construct a type specifier corresponding to a list of two elements such that each element is again a list with three arbitrary elements.

16. Loops

Loops are important elements of MuPAD's programming language.
The following example illustrates the simplest form of a **for** loop:

```
>> for i from 1 to 4 do
       x := i^2;
       print("The square of", i, "is", x)
   end_for:
```

The screen output is:

```
"The square of", 1, "is", 1

"The square of", 2, "is", 4

"The square of", 3, "is", 9

"The square of", 4, "is", 16
```

The loop variable i automatically runs through the values $1, 2, 3, 4$. For each value of i all commands between **do** and **end_for** are executed. There may be arbitrarily many commands, separated by semicolons or colons. *The system does not print the results computed in each loop iteration on the screen, even if you terminate the commands by semicolons.* For that reason we used the **print** command to generate an output in the above example.

The following variant counts backwards. We use the tools from Section 4.11 to make the output look more appealing:

```
>> for j from 4 downto 2 do
       print(Unquoted,
              "The square of ".expr2text(j)." is ".
              expr2text(j^2))
   end_for:
```

The square of 4 is 16

The square of 3 is 9

The square of 2 is 4

You can use the keyword **step** to increment or decrement the loop variable in bigger steps:

```
>> for x from 3 to 8 step 2 do print(x, x^2) end_for:
```

3, 9

5, 25

7, 49

Note that at the end of the iteration with $x = 7$ the value of x is incremented to 9. This exceeds the upper bound 8, and the loop terminates. Here is another variant of the **for** loop:

```
>> for i in [5, 27, y] do print(i, i^2) end_for:
```

5, 25

27, 729

$$
\begin{array}{c}
2 \\
\text{y, y}
\end{array}
$$

The loop variable only runs through the values from the list $[5, 27, y]$. As you can see, such a list may contain symbolic elements such as the variable y.

In a **for** loop a loop variable changes according to fixed rules (typically it is incremented or decremented). The **repeat** loop is a more flexible alternative where you can arbitrarily modify many variables in each step. In the following example we compute the squares of the integers $i = 2, 2^2, 2^4, 2^8, \ldots$ until $i^2 > 100$ holds for the first time:

```
>> x := 2:
```

```
>> repeat
      i := x; x := i^2; print(i, x)
   until x > 100 end_repeat:
```

```
2, 4
```

```
4, 16
```

```
16, 256
```

The system executes the commands between **repeat** and **until** repeatedly. The loop terminates, when the condition stated between **until** and **end_repeat** holds true. In the above example we have $i = 4$ and $x = 16$ at the end of the second step. Hence the third step is executed, and afterwards we have $i = 16, x = 256$. Now the termination condition $x > 100$ is true and the loop terminates.

Another variant in MuPAD is the **while** loop:

```
>> x := 2:
>> while x <= 100 do
      i := x; x := i^2; print(i, x)
   end_while:
```

```
2, 4
```

```
4, 16
```

```
16, 256
```

In a **repeat** loop, the system checks the termination condition *after* each loop iteration. In a **while** loop, this condition is checked *before* each iteration. As soon as the condition evaluates to **FALSE**, the system terminates the **while** loop.

You can use **break** to abort a loop explicitly. Typically this is done within an **if** construction (Chapter 17):

```
>> for i from 3 to 100 do
      print(i);
      if i^2 > 20 then break end_if
   end_for:
```

3

4

5

After a call to **next**, the system skips all commands up to **end_for**. It returns immediately to the beginning of the loop and starts the next iteration with the next value of the loop variable:

```
>> for i from 2 to 5 do
      x := i;
      if i > 3 then next end_if;
      y := i;
      print(x, y)
   end_for:
```

2, 2

3, 3

For $i > 3$ only the first assignment **x:=i** is executed:

```
>> x, y;
```

5, 3

We recall that every MuPAD command returns an object. For a loop this is the return value of the most recently executed command. If you terminate the loop command with a semicolon (and not with a colon as in all of the above examples), then MuPAD displays this value:

```
>> unassign(x): for i from 1 to 3 do x.i := i^2 end_for;
```

9

You may process this value further. In particular, you can assign it to an identifier or use it as the return value of a MuPAD procedure (Chapter 18):

```
>> factorial := proc(n)
     local result;
     begin
       result := 1;
       for i from 2 to n do
         result := result * i
       end_for
     end_proc:
```

The return value of the above procedure is the return value of the **for** loop, which in turn is the value of the last assignment to **result**.

Internally loops are system function calls. For example, MuPAD processes a **for** loop by evaluating the function **_for**:

```
>> _for(i, first_i, last_i, increment, command):
```

This is equivalent to

```
>> for i from first_i to last_i step increment do
       command
     end_for:
```

17. Branching: `if-then-else` and `case`

Branching instructions are an important element of every programming language. Depending on the value or the meaning of variables different commands are executed. The simplest variant in MuPAD is the `if` statement:

```
>> for i from 2 to 4 do
     if isprime(i)
        then print(expr2text(i)." is prime")
        else print(expr2text(i)." is not prime")
     end_if
   end_for:

   "2 is prime"

   "3 is prime"

   "4 is not prime"
```

Here the primality test `isprime(i)` returns either `TRUE` or `FALSE`. If the value is `TRUE`, then the system executes the commands between `then` and `else` (in this case only one `print` command). If it is `FALSE`, then the commands between `else` and `end_if` are executed. The `else` branch is optional:

```
>> for i from 2 to 4 do
     if isprime(i)
        then text := expr2text(i)." is prime";
             print(text)
     end_if
   end_for:
```

"2 is prime"

"3 is prime"

Here the **then** branch comprises two commands separated by a semi-colon (or alternatively a colon). You may nest commands, loops, and branching statements arbitrarily:

```
>> primes := []: evenNumbers := []:
>> for i from 30 to 50 do
      if isprime(i)
         then primes := primes.[i]
         else if testtype(i,Type::Even)
                 then evenNumbers := evenNumbers.[i]
              end_if
      end_if
   end_for:
```

In this example, we inspect the integers between 30 and 50. If we encounter a prime then we append it to the list **primes**. Otherwise we use **testtype** to check whether i is even (see Sections 15.1 and 15.2). In that case, we append i to the list **evenNumbers**. Upon termination the list **primes** contains all prime numbers between 30 and 50, and **evenNumbers** contains all even integers in this range:

```
>> primes, evenNumbers;
```

$$[31, 37, 41, 43, 47], [30, 32, 34, 36, 38, 40, 42, 44,$$

$$46, 48, 50]$$

You can create more complex conditions for the **if** statement by using the Boolean operators **and**, **or**, and **not** (Section 4.10). The following for loop outputs prime twins $[i, i + 2]$. The alternative condition **not** (i>3) yields in addition the pair $[2, 4]$:

```
>> for i from 2 to 100 do
      if (isprime(i) and isprime(i+2)) or not (i>3)
         then print([i,i+2])
      end_if
   end_for:
```

[2, 4]

[3, 5]

[5, 7]

[11, 13]

...

Internally an if statement is just a call of the system function _if:

```
>> _if(condition, command1, command2):
```

is equivalent to

```
>> if condition then command1 else command2 end_if:
```

Thus the commands

```
>> x := 1234567:
>> _if(isprime(x), print("prime"), print("not prime")):
```

yield the output:

```
"not prime"
```

The return value of an if statement—and more generally of any Mu-PAD procedure—is the result of the most recently executed command between if and end_if[1]:

```
>> x := -2: if x > 0 then x else -x end_if;
```

    2

For example, you can use the arrow operator -> (Section 4.12) to implement the absolute value for numbers, as follows:

```
>> Abs := y -> (if y > 0 then y else -y end_if):
```

---
[1] If no command is executed then the result is NIL.

```
>> Abs(-2), Abs(-2/3), Abs(3.5);
```

   2, 2/3, 3.5

As you can see you can use if commands in MuPAD both at interactive level and within procedures. The typical application is in programming MuPAD procedures, where if statements and loops control the flow of the algorithm. A simple example is the above function Abs. You find more examples in Chapter 18.

If you have several nested if .. else if .. constructions, then you can abbreviate this by using the elif statement:

```
>> if condition1 then
      statements1
   elif condition2 then
      statements2
   elif ...
   else
      statements
   end_if:
```

This is equivalent to the following nested if statement:

```
>> if condition1 then
      statements1
   else if condition2 then
         statements2
      else if ...
            else
               statements
            end_if
      end_if
   end_if:
```

A typical application is for type checking within procedures (Chapter 18). The following version of Abs computes the absolute value if the input is an integer, a rational number, a floating-point number, or a complex number, and issues an error message otherwise:

```
>> Abs := proc(y) begin
      if (domtype(y) = DOM_INT) or (domtype(y) = DOM_RAT)
         or (domtype(y) = DOM_FLOAT) then
            if y > 0 then y else -y end_if;
      elif (domtype(y) = DOM_COMPLEX) then
            sqrt(Re(y)^2 + Im(y)^2);
      else "Invalid argument type" end_if:
   end_proc:
>> unassign(x): Abs(-3), Abs(5.0), Abs(1+2*I), Abs(x);
```

$$
3,\ 5.0,\ 5^{1/2},\ \text{"Invalid argument type"}
$$

In our example, we distinguish several cases according to the evaluation of a single expression. We can also implement this by using a `case` statement, which is often easier to read:

```
>> case domtype(y)
   of DOM_INT do
   of DOM_RAT do
   of DOM_FLOAT do
      if y > 0 then y else -y end_if;
      break;
   of DOM_COMPLEX do
      sqrt(Re(y)^2 + Im(y)^2);
      break;
   otherwise
      "Invalid argument type";
   end_case:
```

The keywords `case` and `end_case` indicate the beginning and the end of the statement, respectively. MuPAD evaluates the expression after `case`. If the result matches one of the expressions between `of` and `do`, then the system executes all commands from the first matching `of` on until it encounters either a `break` or the keyword `end_case`. This is in the same style as the `switch` statement in the C programming language. If none of the `of` branches applies and if there is an `otherwise` branch, then the code between `otherwise` and `end_case` is executed. The return value of a `case` statement is the value of the last executed command. We refer to the corresponding help page `?case` for a more detailed description.

As for loops and for the `if` statement, there is a functional equivalent for a `case` statement: the system function `_case`. Internally MuPAD converts the above `case` statement to the following equivalent form:

```
>> _case(domtype(y),
        DOM_INT, NIL,
        DOM_RAT, NIL,
        DOM_FLOAT,
          (if y > 0 then y else -y end_if; break),
        DOM_COMPLEX, (sqrt(Re(y)^2 + Im(y)^2); break),
        "Invalid argument type"):
```

Exercise 17.1: In `if` statements or termination conditions of `while` and `repeat` loops, the system evaluates composite conditions with Boolean operators one after the other. The evaluation routine stops prematurely if it can decide whether the final result is TRUE or FALSE ("lazy evaluation"). Are there problems with the following statements? What happens when the conditions are evaluated?

```
>> A := x/(x-1) > 0: x := 1:
>> (if x <> 1 and A then right else wrong end_if),
   (if x = 1 or A then right else wrong end_if);
```

18. **MuPAD** Procedures

MuPAD provides the essential constructs of a programming language. The user can implement complex algorithms comfortably in MuPAD. Indeed, most of MuPAD's mathematical intelligence is not implemented in C or C++ within the kernel but in MuPAD's programming language at the library level. The programming features are more extensive than in other languages such as C, Pascal, or Fortran, since MuPAD offers more general and more flexible constructs.

We have already presented basic structures such as loops (Chapter 16), branching instructions (Chapter 17), and "simple" functions (Section 4.12).

In this chapter, we regard "programming" as writing complex MuPAD procedures. In principle the user recognizes no differences between "simple functions" generated via -> (Section 4.12) and "more complex procedures" as presented later. Procedures, like functions, return values. Only the way of generating such procedure objects via proc-end_proc is a little more complicated. Procedures provide additional functionality: there is a distinction between local and global variables, you can use arbitrarily many commands in a clear and convenient way, etc.

As soon as a procedure is implemented, you may call it in the form procedureName(arguments) like any other MuPAD function. After executing the implemented algorithm it returns an output value.

You can define and use MuPAD procedures within an interactive session, like any other MuPAD object. However, you often want to these procedures in later sessions again, in particular when they implement more complex algorithms. Then it is useful to write the procedure definition with your favorite text editor, store it in a text file, and read it into a MuPAD session via read (Section 13.2.1). Apart from the evaluation level the MuPAD kernel processes the commands in the file exactly in the same way as if they would have been entered interactively.

18.1 Defining Procedures

The following function, which compares two numbers and returns their maximum, is a simple example of a procedure definition via proc-end_proc:

```
>> Max := proc(a, b) /* comment: maximum of a and b */
        begin
           if a<b then return(b) else return(a) end_if
        end_proc:
```

The text enclosed between /* and */ is a comment[1] and completely ignored by the system. This is a useful tool for documenting the source code when you write the procedure definition in a text file.

The above sample procedure contains an **if** statement as the only command. More realistic procedures contain many commands (separated by colons or semicolons). The command **return** terminates a procedure and passes its argument as output value to the system.

A MuPAD object generated via **proc-end_proc** is of domain type DOM_PROC:

```
>> domtype(Max);
```

```
    DOM_PROC
```

You can decompose and manipulate a procedure like any other MuPAD object. In particular, you may assign it to an identifier, as above. The syntax of the function call is the same as for other MuPAD functions:

```
>> Max(3/7, 0.4);
```

```
    3/7
```

The statements between **begin** and **end_proc** may be arbitrary MuPAD commands. In particular you may call system functions or other procedures from within a procedure. A procedure may even call itself, which is helpful for implementing recursive algorithms. The favorite

---

[1] Alternatively you can start a comment by //, which indicates implicitly that the comment ends at the end of the line. Enclosing comments in #, as was common in earlier MuPAD versions, is no longer valid in versions beyond 1.4.

example for a recursive algorithm is the factorial $n! = 1 \cdot 2 \cdot \ldots \cdot n$ of a nonnegative integer, which may be defined by the rule $n! = n \cdot (n-1)!$ together with the initial condition $0! = 1$. The realization as a recursive procedure might look as follows:

```
>> factorial := proc(n)  begin
     if n = 0
        then return(1)
        else return(n*factorial(n - 1))
     end_if
   end_proc:
>> factorial(10);

   3628800
```

The environment variable MAXDEPTH determines the maximal number of nested procedure calls. Its default value is 500. With this value the above factorial function works only for $n \leq 500$. For larger values it assumes after MAXDEPTH steps that there is an infinite recursion and aborts with an error message. After increasing the value of MAXDEPTH you can compute factorials for larger values.

18.2 The Return Value of a Procedure

When you call a procedure the system executes its body, i.e., the sequence of statements between **begin** and **end_proc**. Every procedure returns some value, either explicitly via **return** or otherwise *the value of the most recently executed command within the procedure*[2]. Thus you can implement the above factorial function without using **return**:

```
>> factorial := proc(n) begin
     if n = 0 then 1 else n*factorial(n - 1) end_if
   end_proc:
```

The **if** statement returns either 1 or $n(n-1)!$, and this is then the output value of the call **factorial(n)**.

As soon as the system encounters a **return** statement it terminates the procedure:

---

[2] If there is no command to execute, then the return value is NIL.

```
>> factorial := proc(n) begin
     if n = 0 then return(1) end_if;
     n*factorial(n - 1)
   end_proc:
```

For $n = 0$, MuPAD does not execute the last statement (the recursive call of n*factorial(n-1)) after returning 1. For $n \neq 0$ the most recently computed value is n*factorial(n-1), which is then the return value of the call factorial(n).

A procedure may return an arbitrary MuPAD object, such as an expression, a sequence, a set, a list, or even a function or a procedure[3]:

```
>> generatePowerFunction := proc(power)
         begin
           x -> (x^power)
         end_proc:
>> f := generatePowerFunction(2);

   x -> (x^2)

>> g := generatePowerFunction(5);

   x -> (x^5)

>> f(a), g(b);

    2   5
   a , b
```

18.3 Returning Symbolic Function Calls

Many system functions return "themselves" as symbolic function calls, if they cannot find a simple representation of the requested result:

```
>> sin(x), max(a, b), int(exp(x^3), x);
```

---

[3] In MuPAD versions beyond 1.4, the example does not work; you have to declare generatePowerFunction with option escape to make it work.

$$sin(x), \ max(a, \ b), \ int(exp(x^3), \ x)$$

You achieve the same behavior in your own procedures, when you encapsulate the procedure name in a hold upon return. The hold (Section 5.2) prevents the function from calling itself recursively and ending up in an infinite recursion. The following function computes the absolute value for numerical inputs (integers, rational numbers, floating-point numbers, and complex numbers). For all other kinds of inputs, it returns itself unevaluated[4]:

```
>> Abs := proc(x) begin
      if testtype(x, NUMERIC) then
         if domtype(x) = DOM_COMPLEX
            then return(sqrt(Re(x)^2 + Im(x)^2))
            else if x >= 0
                     then return(x)
                     else return(-x)
                  end_if
         end_if
      end_if;
      hold(Abs)(x)
   end_proc:
>> Abs(-1), Abs(-2/3), Abs(1.234), Abs(2 + I/3),
   Abs(x + 1);
```

$$1, \ 2/3, \ 1.234, \ \frac{37^{1/2}}{3}, \ Abs(x + 1)$$

A more elegant way is to use the MuPAD object **procname**, which returns the name of the calling procedure. Thus you need not fix a name a priori when defining the procedure:

---

[4] In MuPAD versions beyond 1.4, NUMERIC is replaced by Type::Numeric.

```
>> proc(x) begin
    if testtype(x, NUMERIC) then
        if domtype(x) = DOM_COMPLEX
            then return(sqrt(Re(x)^2 + Im(x)^2))
            else if x >= 0
                    then return(x)
                    else return(-x)
                 end_if
        end_if
    end_if;
    procname(args())
  end_proc:
>> Abs := %: AbsValue := %2: Abs(x + 1), AbsValue(x + 1);

    Abs(x + 1), AbsValue(x + 1)
```

Here we use the expression **args()**, which returns the sequence of arguments passed to the procedure (Section 18.7).

18.4 Local and Global Variables

You can use arbitrary identifiers in procedures. They are also called *global variables*:

```
>> a := b: f := proc() begin a := 1 + a^2 end_proc:
>> f(); f(); f();

    b + 1

       2     2
    (b  + 1)  + 1

        2    2    2
    ((b  + 1)  + 1)  + 1
```

The procedure **f** modifies the value of **a**, which has been set outside the procedure. When the procedure terminates, **a** has a new value, which is again changed by further calls to **f**.

The keyword **local** declares identifiers as *local variables* that are only valid within the procedure:

```
>> a := b: f := proc() local a; begin a := 2 end_proc:
>> f(): a;
```

b

Despite the equal names the assignment a:=2 of the local variable does not affect the value of the global identifier a that has been defined outside the procedure. You can declare an arbitrary number of local variables by specifying a sequence of identifiers after local:

```
>> f := proc(x, y, z) local A, B, C;
       begin
           A:= 1; B:= 2; C:= 3; A*B*C*(x + y + z)
       end_proc:
>> f(A, B, C);
```

6 A + 6 B + 6 C

We recommend to take into account the following rule of thumb:

Using global variables is generally considered bad programming style. Use local variables whenever possible.

The reason for this principle is that procedures implement mathematical functions, which should return a unique output value for a given set of input values. If you use global variables then, depending on their values, the same procedure call may lead to different results:

```
>> a := 1: f := proc(b) begin a := a + 1; a + b end_proc:
>> f(1), f(1), f(1);
```

3, 4, 5

Moreover, a procedure call can change the calling environment in a subtle way by redefining global variables ("side effect"). In more complex programs, this may lead to unwanted effects that are difficult to debug.

We now present a realistic example of a meaningful procedure. If we use arrays of domain type DOM_ARRAY to represent matrices, then we are faced with the problem that there is no direct way in Mu-PAD to perform matrix multiplication with such arrays[5]. The following

---

[5] If you use the data type Dom::Matrix() instead, you can immediately use the standard operators +, -, *, ^, / for arithmetic with matrices (Section 4.15.2).

procedure solves this problem: you can compute the matrix product $C = A \cdot B$ with the simple command `C:=MatrixProduct(A,B)`. We want the procedure to work for arbitrary dimensions of the matrices A and B—provided the result is defined mathematically. If A is an $m \times n$ matrix, then B may be an $n \times r$ matrix, where m, n, r are arbitrary positive integers. The result is the $m \times r$ matrix C with the entries

$$C_{ij} = \sum_{k=1}^{n} A_{ik} B_{kj} , \quad i = 1, \dots, m, \quad j = 1, \dots, r.$$

The multiplication procedure below automatically extracts the dimension parameters m, n, r from the arguments, namely from the 0th operands of the input arrays (Section 4.9). If B is a $q \times r$ matrix with $q \neq n$, then the multiplication is not defined mathematically. In this case the procedure terminates with an error message. For that purpose we employ the function `error`, which aborts the calling procedure and writes the string passed as argument on the screen. We store the result component by component in the local variable `C`. We initialize this variable as an array of dimension $m \times r$, so that the result of our procedure is of the desired data type `DOM_ARRAY`. We might implement the sum over k in the computation of C_{ij} as a loop of the form `for k from 1 to n do` .. Instead, we use the system function `_plus` which returns the sum of its arguments. We generally recommend to use these system functions, if possible, since they work quite efficiently. The return value of the final expression `C` is the output value of `MatrixProduct`:

```
>> MatrixProduct := /* multiplication C=AB of an m x n */
   proc(A, B)          /* matrix A by an n x r Matrix B    */
   local m, n, r, i, j, k, C; /* with arrays A, B of      */
   begin                      /* domain type DOM_ARRAY */
       m := op(A, [0, 2, 2]);
       n := op(A, [0, 3, 2]);
       if n <> op(B, [0, 2, 2]) then
         error("incompatible matrix dimensions")
       end_if;
       r := op(B, [0, 3, 2]);
       C := array(1..m, 1..r);        /* initialization */
```

```
for i from 1 to m do
  for j from 1 to r do
    C[i, j] := _plus(A[i, k]*B[k, j] $ k = 1..n)
  end_for
end_for;
C
end_proc:
```

A general remark about programming style in MuPAD: you should always perform argument checking in procedures meant for interactive use. If you implement a procedure, you usually know which types of inputs are valid (such as DOM_ARRAY in the above example). If somebody passes parameters of the wrong type by mistake, this usually leads to system functions calls with invalid arguments, and your procedure aborts with an error message originating from a system function. In the above example the function op returns the value FAIL when accessing the 0th operand of A or B and one of them is not of type DOM_ARRAY. Then this value is assigned to m,n or r, and the following for loop aborts with an error message, since FAIL is not allowed as a value for the upper endpoint of the loop.

In such a situation it is often difficult to locate the source of the error. However, a worse scenario might even happen: if the procedure does not abort, then the result is likely to be wrong! Thus type checking helps to avoid errors.

In the above example we might add a type check of the form

```
>> if domtype(A) <> DOM_ARRAY or domtype(B) <> DOM_ARRAY
     then error("arguments must be of type DOM_ARRAY")
   end_if;
```

to the procedure body. In Section 18.6 we discuss a simpler type checking concept.

18.5 Subprocedures

Often tasks occur frequently within a procedure and you want to implement them again in the form of a procedure. This structures and simplifies the program code. In many cases, such a procedure is used only from within a single procedure. Then it is reasonable to define this procedure locally as a subprocedure only in the scope of the calling procedure. In MuPAD you can use local variables to implement subprocedures. If you want to make

```
g := proc() begin ... end_proc:
```

a local procedure of

```
f := proc() begin ... end_proc
```

define f as follows:

```
>> f := proc()
   local g;
   begin
      g := proc() begin ... end_proc;    /* subprocedure */

      /* main part of f, which calls g(..): */
      ...
   end_proc:
```

Then g is a *local* procedure of f and you can use it only from within f.

We state a simple example. You can implement matrix multiplication by means of suitable column×row multiplications:

$$
\begin{pmatrix} 2 & 1 \\ 5 & 3 \end{pmatrix} \cdot \begin{pmatrix} 4 & 6 \\ 2 & 3 \end{pmatrix} = \begin{pmatrix} (2,1) \cdot \begin{pmatrix} 4 \\ 2 \end{pmatrix} & (2,1) \cdot \begin{pmatrix} 6 \\ 3 \end{pmatrix} \\ (5,3) \cdot \begin{pmatrix} 4 \\ 2 \end{pmatrix} & (5,3) \cdot \begin{pmatrix} 6 \\ 3 \end{pmatrix} \end{pmatrix} = \begin{pmatrix} 10 & 15 \\ 26 & 39 \end{pmatrix} .
$$

More generally, if we decompose the input matrices into rows and columns, respectively, then

$$
\begin{pmatrix} a_1 \\ \vdots \\ a_m \end{pmatrix} \cdot (b_1, \ldots, b_n) = \begin{pmatrix} a_1 \cdot b_1 & \ldots & a_1 \cdot b_n \\ \vdots & \ddots & \vdots \\ a_m \cdot b_1 & \ldots & a_m \cdot b_n \end{pmatrix} ,
$$

with the inner product

$$
a_i \cdot b_j = \sum_r (a_i)_r \, (b_j)_r .
$$

We now write a procedure MatrixMult that expects arrays A and B of the form array(1..m,1..k) and array(1..k,1..n), respectively, as arguments and returns the $m \times n$ matrix product $A \cdot B$. A call of the subprocedure RowTimesColumn with arguments i, j extracts the ith row and the jth column from the input matrices A and B, respectively, and computes the inner product of the row and the column. The subprocedure uses the arrays A, B as well as the locally declared dimension parameters m, n, and k as "global" variables:

```
>> MatrixMult := proc(A, B)
   local m, n, k, K,      /* local variables */
         RowTimesColumn; /* local subprocedure */
   begin
     /* subprocedure */
     RowTimesColumn := proc(i, j)
     local row, column, r;
     begin
       /* ith row of A: */
       row := array(1..k, [A[i,r] $ r=1..k]);
       /* jth column of B: */
       column := array(1..k, [B[r,j] $ r=1..k]);
       /* row times column */
       _plus(row[r]*column[r] $ r=1..k)
     end_proc;

     /* main part of the procedure MatrixMult: */
     m := op(A, [0, 2, 2]); /* number of rows of A */
     k := op(A, [0, 3, 2]); /* number of columns of A */
     K := op(B, [0, 2, 2]); /* number of rows of B */
     n := op(B, [0, 3, 2]); /* number of columns of B */

     if k <> K then
        error("# of columns of A <> # of rows of B")
     end_if;

     /* matrix A*B: */
     array(1..m, 1..n,
           [[RowTimesColumn(i, j) $ j=1..n] $ i=1..m])
   end_proc:
```

The following example returns the desired result:

```
>> A := array(1..2, 1..2, [[2, 1], [5, 3]]):
>> B := array(1..2, 1..2, [[4, 6], [2, 3]]):
>> MatrixMult(A, B);
```

```
    +-        -+
    |  10, 15  |
    |          |
    |  26, 39  |
    +-        -+
```

Warning: For the reader with programming experience we note that MuPAD 1.4 implements "dynamic scoping" (we refer to the MuPAD manual [MuP 96] for further information). Future MuPAD versions implement "lexical scoping". Thus you should not exploit the dynamic scoping of MuPAD 1.4 in order to ensure compatibility with future versions.

18.6 Type Declaration

As of version 1.4, MuPAD provides easy-to-use type checking for procedure arguments. For example, you can restrict the arguments of the procedure MatrixProduct from Section 18.4 to the domain type DOM_ARRAY as follows:

```
>> MatrixProduct := proc(A:DOM_ARRAY, B:DOM_ARRAY)
                    local m, n, r, i, j, k, C;
                    begin ...
```

If you declare the type of the parameters of a procedure in the form argument:typeSpecifier, then a call of the procedure with parameters of an incompatible type leads to an error message. In the above example, we used the domain type DOM_ARRAY as a type specifier.

We have discussed MuPAD's type concept in Chapter 15. The type Type::NonNegInt from the Type library corresponds to the set of nonnegative integers. If we use it in the following variant of the factorial function

```
>> factorial := proc(n:Type::NonNegInt) begin
     if n = 0
       then return(1)
       else n*factorial(n - 1)
     end_if
   end_proc:
```

then only nonnegative integers are permitted for the argument n:

```
>> factorial(4);
```

> 24

```
>> factorial(4.0);
```

> Error: Wrong type of argument n (type Type::NonNegInt
>   expected) in procedure factorial. Actual
>   argument is 4.000000000.

```
>> factorial(-4);
```

> Error: Wrong type of argument n (type Type::NonNegInt
>   expected) in procedure factorial. Actual
>   argument is -4.

18.7 Procedures with a Variable Number of Arguments

The system function max computes the maximum of its arguments. You may call it with arbitrarily many arguments:

```
>> max(1), max(3/7, 9/20), max(-1, 3, 0, 7, 3/2, 7.5);
```

> 1, 9/20, 7.5

You can implement this behavior in your own procedures as well. The function args returns the arguments passed to the calling procedure:

```
args(0)     : the number of arguments,
args(i)     : the ith argument, 1 ≤ i ≤ args(0),
args(i..j)  : the sequence of arguments from i to j,
                1 ≤ i ≤ j ≤ args(0),
args()      : the sequence args(1),args(2),.. of all
                arguments.
```

The following function simulates the behavior of the system function max:

```
>> Max := proc() local m, i; begin
     m := args(1);
     for i from 2 to args(0) do
         if m < args(i) then m := args(i) end_if
     end_for:
     m
   end_proc:
>> Max(1), Max(3/7, 9/20), Max(-1, 3, 0, 7, 3/2, 7.5);

   1, 9/20, 7.5
```

Here we initialize m with the first argument. Then we test for each of the remaining arguments whether it is greater than m, and if so, replace m by the corresponding argument. Thus m contains the maximum at the end of the loop. Note that if you call Max with only one argument (so that args(0)=1) then the loop for i from 2 to 1 do ... is not executed at all.

You may use both formal parameters and accesses via args in a procedure:

```
>> f := proc(x, y) begin
         if args(0) = 3
             then x^2 + y^3 + args(3)^4
             else x^2 + y^3
         end_if
       end_proc:
>> f(a, b), f(a, b, c);

    2    3    2    3    4
   a  + b , a  + b  + c
```

The following example is a trivial function returning itself symbolically:

```
>> f := proc() begin procname(args()) end_proc:
>> f(a + b + sin(x));
```

```
    f(a + b + sin(x))
```

18.8 Options: the Remember Table

When declaring MuPAD procedures you can specify *options* that affect
the execution of a procedure call. Besides the option hold[6] the option
remember may be of interest to the user. In this section, we take a closer
look at this option and demonstrate its effect with a simple example.
The sequence of Fibonacci numbers is defined by the recursion

$$F_n = F_{n-1} + F_{n-2}, \quad F_0 = 0, \quad F_1 = 1.$$

It is easy to translate this into a MuPAD procedure:

```
>> F := proc(n) begin
            if n < 2 then n else F(n - 1) + F(n - 2) end_if
         end_proc:
>> F(i) $ i = 0..10;
```

```
    0, 1, 1, 2, 3, 5, 8, 13, 21, 34, 55
```

This way of computing F_n is highly inefficient for larger values of n.
To see why, let us trace the recursive calls of F when computing F_4.
You may regard this as a tree structure: F(4) calls F(3) and F(2),
F(3) calls F(2) and F(1), etc.:

---

[6] This option changes the calling semantics for parameters from "call by value" to
"call by name": the arguments are not evaluated. In particular, if you pass an
identifier as argument, then the procedure gets the *name* of the identifier and
not its *value*.

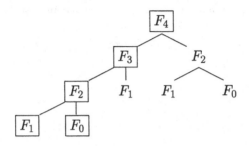

One can show that the call `F(n)` leads to approximately $1.45.. \cdot (1.618..)^n$ calls of `F` for large n. These "costs" grow dramatically fast for increasing values of n:

```
>> time(F(10)), time(F(15)), time(F(20)), time(F(25));

    80, 910, 10290, 113600
```

We recall that the function `time` returns the time in milliseconds used to evaluate its argument.

We see that many calls (such as, for example, `F(1)`) are executed several times. For evaluating `F(4)`, it is sufficient to execute only the boxed function calls `F(0),..,F(4)` in the above figure and to store these values. All other computations of `F(0),F(1),F(2)` are redundant since their results are already known. This is precisely what Mu-PAD does when you declare `F` with the `option remember`:

```
>> F := proc(n)
       /* local x, y; (declare local variables here) */
       option remember;
       begin
          if n < 2 then n else F(n - 1) + F(n - 2) end_if
       end_proc:
```

The system internally creates a remember table for the procedure `F`, which initially is empty. At each call to `F`, MuPAD checks whether there is an entry for the current argument sequence in this table. If this is the case, then the procedure is not executed at all, and the result is taken from the table. If the current arguments do not appear in the table, then the system executes the procedure body as usual and returns its result. Then it appends the argument sequence and the return value to the remember table. This ensures that a procedure is not unnecessarily executed twice with the same arguments.

In the Fibonacci example, the call F(n) now leads to only $n + 1$ calls to compute F(0),..,F(n). In addition, the system searches the remember table $n - 2$ times. However, this happens very quickly. In this example the benefit in speed from using option remember is really dramatic:

```
>> time(F(10)), time(F(15)), time(F(20)), time(F(25)),
   time(F(500));

   0, 10, 0, 10, 390
```

The real running times are so small that the system cannot measure them exactly. This explains the (rounded) times of 0 milliseconds for F(10) and F(20).

> Using option remember in a procedure is promising whenever a procedure is called frequently with the same arguments.

Of course you can simulate this behavior for the Fibonacci numbers directly by computing F_n iteratively instead of recursively and storing already computed values in a table[7]:

```
>> F := proc(n) local i, F; begin
     F[0] := 0: F[1] := 1:
     for i from 2 to n do
         F[i] := F[i - 1] + F[i - 2]
     end_for
   end_proc:
>> time(F(10)), time(F(15)), time(F(20)), time(F(25)),
   time(F(500));

   10, 0, 10, 0, 400
```

The function numlib::fibonacci from the number theory library is yet faster for large arguments.

Warning: The remember mechanism recognizes only previously processed inputs, but does not consider the values of possibly used

---

[7] The following procedure is not properly implemented: what happens when you call F(0)?

global variables. When the values of these global variables change, then the remembered return values are usually wrong. In particular, this is the case for global environment variables such as DIGITS:

```
>> floatexp := proc(x) option remember;
                  begin float(exp(x)) end_proc:
>> DIGITS := 20:
>> floatexp(1); DIGITS := 50: floatexp(1); float(exp(1));
```

2.7182818284590452353

2.7182818284590452353360287471344923927

2.7182818284590452353360287471352662497757 2470936999

Here the system outputs the remembered value of floatexp(1) with higher precision after switching from 20 to 50 DIGITS. Nevertheless this is still the value computed for DIGITS=20; the output only shows all digits that were used *internally*[8] in this computation. The last of the three numbers is the true value of exp(1) computed with 50 digits. It differs from the wrongly remembered value at the 30th decimal digit.

You can explicitly add new entries to the remember table of a procedure. In the following example, f is the function $x \mapsto \sin(x)/x$, which has a removable singularity at $x = 0$. The limit is $f(0) := \lim_{x \to 0} \sin(x)/x = 1$:

```
>> f := proc(x) begin sin(x)/x end_proc: f(0);
```

    Error: division by zero;
              in procedure 'f'

You can easily add the value for $x = 0$:

```
>> f(0) := 1: f(0);
```

1

---

[8] Internally MuPAD uses a certain number of additional "guard digits" exceeding the number of digits requested via DIGITS. However, for the output the system truncates the internally computed value to the desired number of digits.

The assignment `f(0):=1` creates a remember table for `f`, so that a later call of `f(0)` does not try to evaluate the value of $\sin(x)/x$ for $x = 0$. Now you can use `f` without running into danger at $x = 0$ (for example, you can plot it).

Warning: The call:

```
>> unassign(f): f(x) := x^2:
```

does *not* generate the function $f : x \mapsto x^2$, but rather creates a remember table for the identifier `f` with the entry `x^2` *only for the symbolic identifier x*. Any other call to `f` returns a symbolic function call:

```
>> f(x), f(y), f(1);
```

```
    2
  x , f(y), f(1)
```

18.9 Input Parameters

The declared formal arguments of a procedure can be used like additional local variables:

```
>> f := proc(a, b) begin a := 1; a + b end_proc:
>> a := 2: f(a, 1): a;
```

```
  2
```

Modifying `a` within a procedure does not affect the identifier `a` that is used outside the procedure. You should be cautious when you access the calling arguments via `args` (Section 18.7) in a procedure after changing some input parameter. In MuPAD versions up to 1.4, `args` then returns the unchanged input parameter:

```
>> f := proc(a) begin a := 1; a, args(1) end_proc: f(abc);
```

```
  1, abc
```

In later MuPAD versions, **args** returns the modified value, so that f(abc) returns 1,1 in the above example.

In principle arbitrary MuPAD objects may be input parameters of a procedure. Thus you can use sets, lists, expressions, or even procedures and functions:

```
>> p := proc(f) local i; begin
        [f(1), f(2), f(3)]
     end_proc:
>> p(g);

   [g(1), g(2), g(3)]

>> p(proc(x) begin x^2 end_proc);

   [1, 4, 9]

>> p(x -> (x^3));

   [1, 8, 27]
```

In general user-defined procedures evaluate their arguments ("call by value"): when you call f(x) then the procedure f knows only the value of the identifier x. If you declare a procedure with the **option** hold, then the "call by name" scheme is used: the expression or the name of the identifier specified as argument in the call is passed to the procedure. In this case, you can use **context** to obtain the value of the expression:

```
>> f := proc(x) option hold;
        begin x = context(x) end_proc:
>> x := 2: y := 3:
>> f(x), f(y), f(sin(0)), f(sin(0.5)), f(sin(PI));

   x = 2, y = 3, sin(0) = 0, sin(0.5) = 0.4794255386,

      sin(PI) = 0
```

18.10 Evaluation Within Procedures

In Chapter 5, we have discussed MuPAD's evaluation strategy: complete evaluation at interactive level. Thus all identifiers are replaced by their values *recursively* until only symbolic identifiers without a value remain (or the evaluation depth given by the environment variable LEVEL is reached):

```
>> unassign(a, b, c): x := a + b: a := b + 1: b := c: x;

   2 c + 1
```

In contrast, the system performs evaluation within procedures not completely but only with evaluation depth 1. This is similar to internally replacing each identifier by `level(identifier,1)`: every identifier is replaced by its value, but *not recursively*. We recall from Section 5.1 the distinction between the *value* of an identifier (the evaluation at the time of assignment) and its *evaluation* (the "current value", where symbolic identifiers that have been assigned a value in the meantime are replaced by their values as well). In interactive mode, calling an object yields its complete evaluation, while in procedures only the object's value is returned. This explains the difference between the interactive result above and the following result:

```
>> f := proc() begin
          unassign(a, b, c):
          x:= a + b: a:= b + 1: b:= c:
          x
       end_proc:
>> f();

   a + b
```

The reason why two different behaviors are implemented is that the strategy of incomplete evaluation makes the evaluation of procedures faster and increases the efficiency of MuPAD procedures considerably. For a beginner in programming MuPAD, this evaluation concept has its pitfalls. However, after some practice you acquire an appropriate programming style so that you can work with the restricted evaluation depth without problems.

Warning: If you do not work interactively with MuPAD but instead use an editor to write your MuPAD commands in a text file and read them into a MuPAD session via **read** (Section 13.2.1), then these commands are executed within a procedure (namely, **read**). Consequently, the evaluation depth is 1. You can use the system function **level** (Section 5.2) to control the evaluation depth and to enforce complete evaluation if necessary:

```
>> f := proc() begin
         unassign(a, b, c):
         x:= a + b: a:= b + 1: b:= c:
         level(x)
      end_proc:
>> f();

   2 c + 1
```

Warning: In MuPAD versions beyond 1.4, **level** does not evaluate local variables. Moreover you cannot use local variables as symbolic identifiers; you must initialize them before use. The following procedure, in which a symbolic identifier x is passed to the integration function, is no longer valid in MuPAD versions beyond 1.4:

```
>> f := proc(n) local x; begin int(exp(x^n), x) end_proc:
```

You can pass the name of the integration variable as additional argument to the procedure. Thus the following variants are valid:

```
>> f := proc(n, x) begin int(exp(x^n), x) end_proc:
>> f := proc(n, x) local y; begin
         y := x; int(exp(y^n), y) end_proc:
```

If you need symbolic identifiers for intermediate results, then you can generate an identifier without a value via **genident()** and assign it to a local variable.

18.11 Function Environments

MuPAD provides a variety of tools for handling built-in mathematical standard functions such as **sin**, **cos**, **exp**. These tools implement the mathematical properties of these functions. Typical examples are the **float** conversion routine, the differentiation function **diff**, or the function **expand**, which you use to manipulate expressions:

```
>> float(sin(1)), diff(sin(x), x, x, x),
   expand(sin(x + 1));
```

```
0.8414709848, -cos(x), cos(x) sin(1) + sin(x) cos(1)
```

In this sense, the mathematical knowledge about the standard functions is distributed over several system functions: the function **float** has to know how to compute numerical approximations to the sine function, **diff** must know its derivative, and **expand** has to know the addition theorems of the trigonometric functions.

You can invent arbitrary new functions as symbolic names or implement them in procedures. How can you pass the knowledge about the mathematical meaning and the rules of manipulation for the new functions to the other system functions? For example, how can you tell the differentiation routine **diff** what the derivative of your newly created function is? If the function is composed of standard functions known to MuPAD, such as, for example, $f : x \mapsto x \sin(x)$, then this is no problem. The call

```
>> f := x -> (x*sin(x)): diff(f(x), x);
```

```
sin(x) + x cos(x)
```

immediately yields the desired answer. However, there are often situations where the newly implemented function cannot be composed from standard objects. An example from mathematical physics is the "Airy" function Ai, which is uniquely defined by the differential equation

$$\frac{d^2}{dx^2} Ai(x) = x \, Ai(x)$$

together with certain initial conditions for the values of the function and its derivative at $x = 0$. This is a "special function" like sin, cos, ln, etc., but it cannot be represented in terms of these standard functions. Nevertheless it satisfies certain rules. The differential equation implies that you can express the second derivative in terms of the function, the third derivative in terms of the function and its first derivative, and so on. We can compute symbolically in this sense: if the MuPAD identifiers `Ai` and `Ai1` represent the Airy function and its first derivative, respectively, then we would like that MuPAD automatically expresses the higher derivatives in terms of `Ai` and `Ai1`.

Thus our goal is to hand the rules of manipulation (floating-point approximation, differentiation, etc.) for *symbolic* function identifiers to the MuPAD functions `float`, `diff`, etc. This is the actual challenge when you "implement a new mathematical function in MuPAD": to distribute the knowledge about the mathematical meaning of the symbol to MuPAD's standard tools. Indeed this is a necessary task: for example, if you want to differentiate a more complex expression containing both the new function and some standard functions, then this is only possible via the system's differentiation routine. Thus the latter has to learn how to handle the new symbols.

For that purpose MuPAD provides the domain type `DOM_FUNC_ENV` (short for: function environment). Indeed, all built-in mathematical standard functions are of this type in order to enable `float`, `diff`, `expand`, etc. to handle them:

```
>> domtype(sin);
```

```
DOM_FUNC_ENV
```

You can call a function environment like any "normal" function or procedure:

```
>> sin(1.7);
```

```
0.9916648104
```

A function environment consists of three operands. The first operand is a procedure that computes the return value of a function call. The second operand is a procedure for printing a symbolic function call on the screen. The third operand is a table containing information how the system functions `float`, `diff`, `expand`, etc. should handle symbolic function calls.

You can look at the procedure for evaluating a function call by `op(.,1)`. Alternatively the function `expose` prints the source code:

```
>> expose(sin);
```

```
proc(x)
  name sin;
  local f;
begin
  if x::sin <> FAIL then
  ...
end_proc
```

In what follows we demonstrate the implementation of the special Airy function Ai defined by the differential equation $Ai''(x) = x\,Ai(x)$ and the initial conditions[9] $Ai(0) = 0$ and $Ai'(0) = 1$. We represent the function and its first derivative by the identifiers Ai and Ai1, respectively. For a symbolic call Ai1(x) we want the system to print Ai'(x) on the screen.

First we define functions Ai and Ai1 that handle the evaluation:

```
>> Ai := proc(x) begin
          if x = 0 then 0 else procname(x) end_if
        end_proc:
>> Ai1 := proc(x) begin
          if x = 0 then 1 else procname(x) end_if
        end_proc:
```

Since the values of the functions are known only at the origin, we use procname (Section 18.3) to return the symbolic expressions Ai(x) and Ai1(x), respectively, for all arguments $x \neq 0$. This yields:

```
>> Ai(0), Ai(1), Ai(x + y), Ai1(0), Ai1(1), Ai1(x + y);

   0, Ai(1), Ai(x + y), 1, Ai1(1), Ai1(x + y)
```

You generate a new function environment by means of func_env[10]:

```
>> Ai := func_env(Ai, NIL, NIL):
>> Ai1Output := proc(f) begin "Ai'(".expr2text(op(f)).")"
               end_proc:
```

---

[9] This is not the usual notation in the theory of special functions, where Ai is a solution of the differential equation for a different set of initial conditions. We use the above initial conditions since they are simpler.

[10] This function is renamed to funcenv in MuPAD versions beyond 1.4.

```
>> Ai1 := func_env(Ai1, Ai1Output, NIL):
```

These commands convert the procedures `Ai` and `Ai1` to function environments. The first arguments are the procedures `Ai` and `Ai1`, respectively, for evaluation. The second argument of `func_env` is the procedure for screen output. We do not specify such a procedure for `Ai`, and hence the system prints any remaining symbolic function call `Ai(x)` in the standard way on the screen. We want that a symbolic expression `Ai1(x)` is displayed as `Ai'(x)`. This is achieved by the second argument of `func_env`, which you should interpret as a conversion routine. On input `Ai1(x)` it returns the MuPAD object to print on the screen instead of `Ai1(x)`. The argument `f`, which represents `Ai1(x)`, is converted to the string `Ai'(x)` (note that `x=op(f)` for `f=Ai1(x)`). The system outputs this string instead of `Ai1(x)` on the screen[11]:

```
>> Ai(0), Ai(1), Ai(x + y), Ai1(0), Ai1(1), Ai1(x + y);
```

```
    0, Ai(1), Ai(x + y), 1, Ai'(1), Ai'(x + y)
```

The third argument to `func_env` is a table of *function attributes*. It tells the system functions `float`, `diff`, `expand`, etc. how to handle symbolic calls of the form `Ai(x)` and `Ai1(x)`. In the example above we do not provide any function attributes. Hence the system functions do not know yet how to proceed and by default return the expression or themselves, respectively, symbolically:

```
>> float(Ai(1)), expand(Ai1(x + y)),
   diff(Ai(x), x), diff(Ai1(x), x);
```

```
    Ai(1), Ai'(x + y), diff(Ai(x), x), diff(Ai'(x), x)
```

We now use the function `funcattr`[12], which adds new entries to the attribute table, to set some function attributes. In the example below the second argument `"diff"` is for the differentiation routine `diff`:

---

[11] Instead of the output routine above you can also specify the kernel routine `built_in(1101,...)` which is responsible for screen output. Its last argument is a string containing the symbolic function name as to be returned:

```
>> Ai1 := func_env(Ai1, built_in(1101, 0, NIL, "Ai'"), NIL):
```

[12] This function is renamed to `slot` in MuPAD versions beyond 1.4.

```
>> Ai := funcattr(Ai, "diff",
                proc(f,x) begin
                        Ai1(op(f))*diff(op(f),x)
                end_proc):
```

This tells `diff` that `diff(f,x)` with a symbolic function call `f=Ai(y)`, where `y` depends on `x`, should apply the procedure passed as third argument to `funcattr`. The well-known chain rule yields

$$\frac{d}{dx}Ai(y) = Ai'(y)\,\frac{dy}{dx}\,.$$

The specified procedure implements this rule, where the inner function in the expression `f=Ai(y)` is given by `y=op(f)`. Since the symbol `Ai1` represents the derivative of Ai, we use it instead of Ai' in the specified procedure. *Now MuPAD knows that the derivative of the function represented by the identifier Ai1 is the function represented by the identifier Ai1.* We have already implemented the screen output of `Ai1` as `Ai'`:

```
>> diff(Ai(z), z), diff(Ai(y(x)), x),
   diff(Ai(x*sin(x)), x);

   Ai'(z), Ai'(y(x)) diff(y(x), x),

   Ai'(x*sin(x)) (sin(x) + x cos(x))
```

The last step is to tell `diff` how to evaluate a call `diff(f,x)` when `f=Ai1(y)` is a symbolic function call. The defining differential equation implies that

$$\frac{d}{dx}Ai'(y) = Ai''(y)\,\frac{dy}{dx} = y\,Ai(y)\,\frac{dy}{dx}\,.$$

We set the `"diff"` attribute of `Ai1` as follows:

```
>> proc(f, x) local y; begin
     y := op(f); y*Ai(y)*diff(y, x)
   end_proc:
>> Ai1 := funcattr(Ai1, "diff", %):
```

As far as `diff` is concerned, the implementation of the Airy function `Ai` is now complete:

```
>> diff(Ai(x), x, x);

   x Ai(x)

>> diff(Ai1(2*x + 3), x, x, x);

                              2
    16 Ai'(x*2 + 3) + 8 (2 x + 3)  Ai(2 x + 3)
```

As an application, we now want MuPAD to compute the first terms of the Taylor expansion of the Airy function around $x = 0$. We can use the function **taylor** since it calls **diff** internally:

```
>> taylor(Ai(x), x = 0, 12);

                          3          4          6
                         x  Ai(0)   x  Ai'(0)  x  Ai(0)
    Ai(0) + x Ai'(0) + -------- + --------- + -------- +
                           6          12         180


        7           9           10
       x  Ai'(0)   x  Ai(0)    x   Ai'(0)        12
     --------- + -------- + ---------- + O(x  )
          504        12960       45360
```

Here the series expansion routine does not evaluate completely (this changes in MuPAD versions beyond 1.4). Only an additional evaluation step takes the known values $Ai(0) = 0$ and $Ai'(0) = 1$ into account. To this end, we convert the series expansion (of domain type **Puiseux**) to a usual MuPAD sum via **expr**. This call leads to an evaluation:

```
>> t := expr(%): t;

            4     7      10
           x     x      x
     x + -- + --- + -----
           12    504    45360
```

Exercise 18.1: Extend the definition of the Airy function in such a way that the MuPAD object `Ai(x,a,b)` represents the solution of the differential equation

$$\frac{d^2}{dx^2}\, Ai(x) = x\, Ai(x)$$

together with the initial conditions $Ai(0) = a$, $Ai'(0) = b$. The expression `Ai1(x,a,b)` is to represent its derivative with respect to x. It is to be printed on the screen as `Ai'(x,a,b)`. Compute the 10th derivative of Ai for $a = 1$ and $b = 0$. Compute the first terms of the Taylor expansion of Ai around $x = 0$ with arbitrary symbolic values a and b.

Exercise 18.2: Implement an absolute value function `Abs` as a function environment. The call `Abs(x)` should return the absolute value for real numbers x of domain type `DOM_INT`, `DOM_RAT`, or `DOM_FLOAT`. For all other arguments, the symbolic output `|x|` should appear on the screen. The absolute value is differentiable on $\mathbb{R}\backslash\{0\}$. Its derivative is

$$\frac{d\,|y|}{d\,x} = \frac{|y|}{y}\,\frac{dy}{dx}\ .$$

Set the `"diff"` attribute accordingly and compute the derivative of `Abs(x^3)`. Compare your result to the corresponding derivative of the system function `abs`.

18.12 A Programming Example: Differentiation

In this section, we discuss a simple example demonstrating the typical functioning of a symbolic MuPAD procedure. We implement a symbolic differentiation routine that computes the derivatives of algebraic expressions composed of additions, multiplications, exponentiations, some mathematical functions (exp, ln, sin, cos, ..), constants, and symbolic identifiers.

This example is only for illustration purposes since MuPAD already provides such a function: the `diff` routine. This function is implemented in the MuPAD kernel and therefore very fast. Thus a user-defined function that is written in MuPAD's programming language can hardly achieve the efficiency of `diff`.

The following algebraic differentiation rules are valid for the class of expressions that we consider:

1) $\dfrac{df}{dx} = 0$, if f does not depend on x,

2) $\dfrac{dx}{dx} = 1$,

3) $\dfrac{d\,(f+g)}{dx} = \dfrac{df}{dx} + \dfrac{dg}{dx}$ (linearity),

4) $\dfrac{d\,a\,b}{dx} = \dfrac{da}{dx}\,b + a\,\dfrac{db}{dx}$ (product rule),

5) $\dfrac{d\,a^b}{dx} = \dfrac{d}{dx}e^{b\,\ln(a)} = e^{b\,\ln(a)}\,\dfrac{d}{dx}\,(b\,\ln(a))$

$\quad = a^b\,\ln(a)\,\dfrac{db}{dx} + a^{b-1}\,b\,\dfrac{da}{dx}$,

6) $\dfrac{d}{dx}\,F(y(x)) = F'(y)\,\dfrac{dy}{dx}$ (chain rule).

Moreover, for some functions F the derivative is known, and we want to take this into account in our implementation. For an unknown function F, we return the symbolic function call of the differentiation routine.

The procedure in Table 18.1 implements the above properties in the stated order. Its calling syntax is Diff(expression,identifier). We have included in (0) an automatic type check of the second argument, which must be a symbolic identifier of domain type DOM_IDENT. In (1) the MuPAD function has checks whether the expression f to be differentiated depends on x. Linearity of differentiation is implemented in (3) by means of the MuPAD function map:

```
>> map(f1(x) + f2(x) + f3(x), Diff, x);

    Diff(f1(x), x) + Diff(f2(x), x) + Diff(f2(x), x)
```

In (4) we decompose a product expression $f = f_1 \cdot f_2 \cdots$: the command a:=op(f,1) determines the first factor $a = f_1$, then subsop(f,1=1) replaces this factor by 1, such that b assumes the value $f_2 \cdot f_3 \cdots$. Then we call Diff(a,x) and Diff(b,x). If $b = f_2 \cdot f_3 \cdots$ is itself a product, then this leads to another execution of step (4) at the next recursion level. In this way (4) handles products of arbitrary length.

```
>> Diff := proc(f, x:DOM_IDENT)                                  //0)
     local a, b, F, y; begin
       if not has(f, x) then return(0) end_if;                   //1)
       if f = x then return(1) end_if;                           //2)
       if type(f) = "_plus" then return(map(f, Diff, x)) end_if; //3)
       if type(f) = "_mult" then
         a := op(f, 1); b := subsop(f, 1 = 1);
         return(Diff(a, x)*b + a*Diff(b, x))                     //4)
       end_if;
       if type(f) = "_power" then
         a := op(f, 1); b := op(f, 2);
         return(f*ln(a)*Diff(b, x) + a^(b - 1)*b*Diff(a, x))     //5)
       end_if;
       if op(f, 0) <> FAIL then
         F := op(f, 0); y := op(f, 1);                           //6)
         if F = hold(exp) then return( exp(y)*Diff(y, x)) end_if; //6)
         if F = hold(ln)  then return(  1/y  *Diff(y, x)) end_if; //6)
         if F = hold(sin) then return( cos(y)*Diff(y, x)) end_if; //6)
         if F = hold(cos) then return(-sin(y)*Diff(y, x)) end_if; //6)
         /* ... specify further known functions here ... */
       end_if;
       procname(args())                                         //7)
     end_proc:
```

Table 18.1. A symbolic differentiation routine

Step (5) differentiates powers. For $f = a^b$ the call op(f,1) returns the base a and op(f,2) the exponent b. In particular, this covers all monomial expressions of the form $f = x^n$ for constant n. The recursive calls to Diff for $a = x$ and $b = n$ then yield Diff(a,x)=1 and Diff(b,x)=0, respectively, and the expression returned in (5) simplifies to the correct result $x^{n-1} n$.

If the expression f is a symbolic function call of the form $f = F(y)$, then we extract the "outer" function F in (6) via F:=op(f,0) (otherwise F gets the value FAIL in general). Next we handle the case where F is a function with one argument y and extract the "inner" function by y:=op(f,1). If F is the name of a function with known derivative (such as $F = \exp, \ln, \sin, \cos$), then we apply the chain rule. It is easy to extend this list of functions F with known derivatives. In particular, you can add a formula for differentiating symbolic expressions of the form int(..). Extensions to functions F with more than one argument are also possible.

Finally, step (7) returns Diff(f,x) symbolically if no simplifications of the expression f happen in steps (1) through (6).

Diff's mode of operation is adopted from the system function `diff`. Compare the following results to those returned by `diff`:

```
>> Diff(x*ln(x + 1/x), x);
```

```
                    /     1  \
                  x | 1 - -- |
                    |     2  |
   /     1 \        \     x  /
  ln| x + - | + ------------
   \    x  /         1
                   x + -
                       x
```

```
>> Diff(f(x)*sin(x^2), x);
```

```
          2                         2
  2 x f(x) cos(x ) + Diff(f(x), x) sin(x )
```

18.13 Programming Exercises

Exercise 18.3: Write a short procedure `date` that takes three integers `month`, `day`, `year` as input and prints the date in the usual way. For example, the call `date(5,3,1990)` should yield the screen output 5/3/1990.

Exercise 18.4: We define the function $f : \mathbb{N} \to \mathbb{N}$ by

$$f(x) = \begin{cases} 3x + 1 & \text{for odd } x, \\ x/2 & \text{for even } x. \end{cases}$$

The "$(3x + 1)$ problem" asks whether for an arbitrary initial value $x_0 \in \mathbb{N}$, the sequence recursively defined by $x_{i+1} := f(x_i)$ contains the value 1. Write a program that on input x_0 returns the smallest index i with $x_i = 1$.

Exercise 18.5: Implement a function Gcd to compute the greatest common divisor of two positive integers. Of course, you should not use the system functions gcd and igcd. Hint: the Euclidean Algorithm for computing the gcd is based on the observation

$$\gcd(a,\ b)\ =\ \gcd(a\ mod\ b,\ b)\ =\ \gcd(b,\ a\ mod\ b)$$

and the facts $\gcd(0,\ b) = \gcd(b,\ 0) = b$.

Exercise 18.6: Implement a function Quadrature. For a function f and a MuPAD list X of numerical values

$$x_0 < x_1 < \ldots < x_n$$

the call Quadrature(f,X) should compute a numerical approximation of the integral

$$\int_{x_0}^{x_n} f(x)\,dx\ \approx\ \sum_{i=0}^{n-1}(x_{i+1} - x_i)\,f(x_i).$$

Exercise 18.7: Newton's method for numerically finding a root of a function $f : \mathbb{R} \mapsto \mathbb{R}$ employs the iteration $x_{i+1} = F(x_i)$, where $F(x) = x - f(x)/f'(x)$. Write a procedure Newton. The call Newton(f,x0,n), with an expression f, should return the first $n + 1$ elements $x_0, \ldots, x_n$ of the Newton sequence.

Exercise 18.8: The *Sierpinski triangle* is a well-known fractal. We define a variant of it as follows. The Sierpinski triangle is the set of all points $(x, y) \in \mathbb{N} \times \mathbb{N}$ with the following property: there exists at least one position in the binary expansions of x and y such that both have a 1-bit at this position. Write a program Sierpinski that on input xmax,ymax plots the set of all such points with integral coordinates in the range $1 \leq x \leq$ xmax, $1 \leq y \leq$ ymax. Hint: the function numlib::g_adic computes the binary expansion of an integer. The function plot2d with the option Mode=List plots lists of points.

Exercise 18.9: A *logical formula* is composed of identifiers and the operators **and**, **or**, and **not**. For example:

```
>> formula := (x and y) or
              ( (y or z) and (not x) and y and z );
```

Such a formula is called *satisfiable* if it is possible to assign the values TRUE and FALSE to all identifiers in such a way that the formula can be evaluated to TRUE. Write a program that checks whether an arbitrary logical formula is satisfiable.

Solutions to Exercises

Exercise 2.1: The help page `?diff` tells you how to compute higher order derivatives:

```
>> diff(sin(x^2), x, x, x, x, x);
```

$$32\, x^5\, \cos(x^2) - 120\, x\, \cos(x^2) + 160\, x^3\, \sin(x^2)$$

You can also use the longer command `diff(diff(.., x), x)`.

Exercise 2.2: The exact representations are:

```
>> sqrt(27) - 2*sqrt(3), cos(PI/8);
```

$$3^{1/2}\ ,\quad \frac{1/2\ (2^{1/2} + 2)^{1/2}}{2}$$

The numerical approximations are:

```
>> DIGITS := 5:
>> float(sqrt(27) - 2*sqrt(3)), float(cos(PI/8));
```

```
1.732, 0.92388
```

They are correct to within 5 digits: the fifth digit of the first number is 0 and suppressed in the output.

Exercise 2.3:

```
>> expand((x^2 + y)^5);
```

$$y^5 + x^{10} + 5 x^8 y + 5 x^2 y^4 + 10 x^4 y^3 + 10 x^6 y^2$$

Exercise 2.4:

```
>> normal((x^2 - 1)/(x + 1));
```

```
  x - 1
```

Exercise 2.5: You can plot the singular function $f(x) = 1/\sin(x)$ on the interval $[1, 10]$ without any problems since the (equidistant) graphical sample points in this interval contain no singularities:

```
>> plotfunc(1/sin(x), x = 1..10);
```

However, on the interval $[0, 10]$, you run into problems. The boundaries belong to the sample points, and an error occurs when the system tries to evaluate $1/\sin(0) = 1/0$ at the left boundary.

Exercise 2.6: MuPAD immediately returns the claimed limits:

```
>> limit(sin(x)/x, x = 0),
   limit((1 - cos(x))/x, x = 0),
   limit(ln(x), x = 0, Right);

   1, 0, -infinity
>> limit(x^sin(x), x = 0),
   limit((1 + 1/x)^x, x = infinity),
   limit(ln(x)/exp(x), x = infinity);

   1, exp(1), 0
>> limit(x^ln(x), x = 0),
   limit((1 + PI/x)^x, x = infinity),
   limit(2/(1 + exp(-1/x)), x = 0, Left);

   infinity, exp(PI), 0
```

The result **undefined** denotes a non-existing limit:

```
>> limit(sin(x)^(1/x), x = 0);

   undefined
```

Exercise 2.7: You obtain the first result in the desired form by factoring:

```
>> sum(k^2 + k + 1 , k = 1..n): % = Factor(%);
```

$$\frac{5n}{3} + n^2 + \frac{n^3}{3} = \frac{n(3n^2 + n + 5)}{3}$$

```
>> sum((2*k - 3)/((k + 1)*(k + 2)*(k + 3)),
       k = 0..infinity);

   -1/4

>> sum(k/(k - 1)^2/(k + 1)^2, k = 2..infinity);

   5/16
```

Exercise 2.8:

```
>> A := Dom::Matrix()([[1,2,3], [4,5,6], [7,8,0]]):
>> B := Dom::Matrix()([[1,1,0], [0,0,1], [0,1,0]]):
>> 2*(A + B), A*B;
```

$$\left(\begin{array}{ccc} 4, & 6, & 6 \\ 8, & 10, & 14 \\ 14, & 18, & 0 \end{array} \right), \left(\begin{array}{ccc} 1, & 4, & 2 \\ 4, & 10, & 5 \\ 7, & 7, & 8 \end{array} \right)$$

```
>> (A - B)^(-1);
```

```
+-                        -+
|  -5/2,   3/2,  -5/7  |
|                         |
|   5/2,  -3/2,   6/7  |
|                         |
|  -1/2,   1/2,  -2/7  |
+-                        -+
```

Exercise 2.9: a) The function `numlib::mersenne` returns a list of values for p yielding 37 of the 38 currently known Mersenne primes that have been found on supercomputers. The actual computation for $1 < p \leq 1000$ can be easily performed in MuPAD:

```
>> select([$1..1000], isprime):
>> select(%, p -> (isprime(2^p - 1)));
```

After some time you obtain the desired list of values of p:

```
[2, 3, 5, 7, 13, 17, 19, 31, 61, 89, 107, 127, 521,

   607]
```

The corresponding Mersenne primes are:

```
>> map(%, p -> (2^p-1));
```

```
[3, 7, 31, 127, 8191, 131071, 524287, 2147483647,

   2305843009213693951, 618970019642690137449562111,

   162259276829213363391578010288127, ... ]
```

b) Depending on your computer's speed you can test only the first 11 or 12 Fermat numbers in a reasonable amount of time. Note that the 12th Fermat number already has 1234 decimal digits.

```
>> Fermat := n -> (2^(2^n) + 1): isprime(Fermat(10));
```

    FALSE

The only known Fermat primes are the first four Fermat numbers.
Indeed, if MuPAD tests the first 12 Fermat numbers, then after some
time it returns the following four values:

```
>> select([Fermat(i) $ i= 1..12], isprime);
```

    [5, 17, 257, 65537]

Exercise 4.1: The first operand of a power is the base, the second is
the exponent. The first and second operand of an equation is the left
and the right hand side, respectively. The operands of a function call
are its arguments:

```
>> op(a^b, 1), op(a^b, 2);
```

    a, b

```
>> op(a=b, 1), op(a=b, 2);
```

    a, b

```
>> op(f(a,b), 1), op(f(a,b), 2);
```

    a, b

Exercise 4.2: The set with the two equations is op(set,1). Its second
operand is the equation y=.., whose second operand is the right hand
side:

```
>> set := solve({x+sin(3)*y = exp(a),
                 y-sin(3)*y = exp(-a)}, {x,y});
```

```
    { {              sin(3) exp(-a)        exp(-a)   } }
    { { x = exp(a) - --------------, y = ---------- } }
    { {                1 - sin(3)         1 - sin(3) } }
```

```
>> y := op(set, [1, 2, 2]);
```

$$\frac{\exp(-a)}{1 - \sin(3)}$$

Use `assign(op(set))` to perform assignments of both unknowns x and y simultaneously.

Exercise 4.3: If at least one number in a numerical expression is a floating- point number, then the result is a floating-point number:

```
>> 1/3 + 1/3 + 1/3, 1.0/3 + 1/3 + 1/3;
```

```
1, 1.0
```

Exercise 4.4: You obtain the desired floating-point numbers immediately:

```
>> float(PI^(PI^PI)), float(exp(PI*sqrt(163)/3));
```

```
1340164183025859352.0, 640320.0
```

Note that only the first 10 digits of these values are reliable since this is the default precision. Indeed, for larger values of DIGITS you find:

```
>> DIGITS := 100:
>> float(PI^(PI^PI)), float(exp(PI*sqrt(163)/3));
```

```
1340164183006357435.2974491296401314150993749745734991\
237787927516586034092619094068148269472611301142,
```

```
640320.0000000006048637350490160394717418188185394757\
714857603665918194652218258286942536340815822646
```

We compute 235 decimal digits of PI to obtain the correct 234th digit after the decimal point. After setting DIGITS:=235 the result is the last shown digit of float(PI). A more elegant way is to multiply by 10^{234}. Then the desired digit is the first digit before the decimal point, and we obtain it by truncating the digits after the decimal point:

```
>> DIGITS := 235: trunc(10^234*PI) - 10*trunc(10^233*PI);
```

    6

Exercise 4.5: a) Internally, MuPAD computes exactly with some additional digits not shown in the output.

```
>> DIGITS := 10: x := 10^50/3.0; floor(x);
```

    3.333333333e49

    33333333333333333333283774257264866876160389152768 0

b) After increasing DIGITS MuPAD displays the additional digits. However, not all of them are correct:

```
>> DIGITS := 40: x;
```

    3.333333333333333333261053188e49

Restart the computation with the increased value of DIGITS to obtain the desired precision:

```
>> DIGITS := 40: x := 10^50/3.0;
```

    3.333333333333333333333333333333333333333e49

Exercise 4.6: The names caution!-!, x-y, and Jack&Jill are invalid since they contain the special characters !, -, and &, respectively. Since an identifier's name must not start with a number, 2x is not valid either. The names diff and exp are valid names of identifiers. However, you cannot assign values to them since they are protected names of MuPAD functions.

Exercise 4.7: We use the sequence operator $ (Section 4.5) to generate the set of equations and the set of unknowns. Then a call to solve returns a set of simpler equations:

```
>> equations := {(x.i + x.(i+1) = 1) $ i = 1..19,
                  x20 = PI}:
>> unknowns := {x.i $ i = 1..20}:
>> solutions := solve(equations, unknowns);
```

$$\{\{x17 = 1 - PI, \ x19 = 1 - PI, \ x2 = PI, \ x10 = PI,$$

$$x4 = PI, \ x20 = PI, \ x12 = PI, \ x1 = 1 - PI, \ x6 = PI,$$

$$x14 = PI, \ x3 = 1 - PI, \ x8 = PI, \ x11 = 1 - PI,$$

$$x16 = PI, \ x5 = 1 - PI, \ x13 = 1 - PI, \ x18 = PI,$$

$$x7 = 1 - PI, \ x15 = 1 - PI, \ x9 = 1 - PI\}\}$$

We use the function **assign** to assign the computed values to the identifiers:

```
>> assign(op(solutions, 1)): x1, x2, x3, x4, x5, x6;
```

$$1 - PI, \ PI, \ 1 - PI, \ PI, \ 1 - PI, \ PI$$

Exercise 4.8: MuPAD stores the expression a^b-sin(a/b) in the form a^b+(-1)*sin(a*b^(-1)). Its expression tree is:

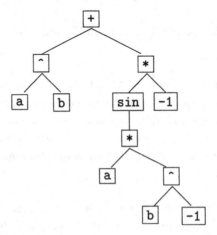

Exercise 4.9: We observe that:

```
>> op(2/3); op(x/3);
```

    2, 3

    x, 1/3

The reason is that 2/3 is of domain type DOM_RAT, whose operands are the numerator and the denominator. The domain type of the symbolic expression x/3 is DOM_EXPR and its internal representation is x*(1/3). The situation is similar for 1+2*I and x+2*I:

```
>> op(1 + 2*I); op(x + 2*I);
```

    1, 2

    x, 2 I

The first object is of domain type DOM_COMPLEX. Its operands are the real and the imaginary part. The operands of the symbolic expression x+2*I are the first and the second term of the sum.

Exercise 4.10: The expression tree of condition = (not a) and (b or c) is:

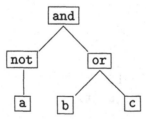

Thus op(condition,1) = not a and op(condition,2) = b or c. We obtain the atoms a, b, c as follows:

```
>> op(condition, [1,1]), op(condition, [2,1]),
   op(condition, [2,2]);
```

    a, b, c

Exercise 4.11: You can use both the assignment function `_assign` presented in Section 4.3 and the assignment operator `:=`.

```
>> _assign(x.i, i) $ i = 1..100:
>> (x.i := i) $ i = 1..100:
```

You can also pass a set of assignment equations to the `assign` function:

```
>> assign({x.i = i $ i = 1..100}):
```

Exercise 4.12: Since a sequence is a valid argument of the sequence operator, you may achieve the desired result as follows:

```
>> (x.i $ i ) $ i = 1..10;
```

```
   x1, x2, x2, x3, x3, x3, x4, x4, x4, x4, ...
```

Exercise 4.13: We use the addition function `_plus` and generate its argument sequence via `$`:

```
>> _plus(((i+j)^(-1) $ j = 1.. i) $ i=1..10);
```

```
   1464232069/232792560
```

Exercise 4.14:

```
>> L1 := [a, b, c, d]: L2 := [1, 2, 3, 4]:
>> L1.L2, zip( L1, L2, _mult);
```

```
   [a, b, c, d, 1, 2, 3, 4]  ,  [a, 2 b, 3 c, 4 d]
```

Exercise 4.15: The function _mult multiplies its arguments:

```
>> map([1, x, 2], _mult, multiplier);

   [multiplier, x multiplier, 2 multiplier]
```

We use map to apply the function L -> map(L,_mult,multiplier) to a nested list:

```
>> L := [[1, x, 2], [PI], [2/3, 1]]:
   map(L, map, _mult, 2);

   [[2, 2 x, 4], [2 PI], [4/3, 2]]
```

Exercise 4.16: For:

```
>> X := [x1, x2, x3]: Y := [y1, y2, y3]:
```

the products are given immediately by

```
>> _plus(op(zip(X, Y, _mult)));

   x1 y1 + x2 y2 + x3 y3
```

The following function f multiplies each element of the list Y by its input parameter x and returns the resulting list:

```
>> f := x -> (map(Y, _mult, x)):
```

The next command replaces each element of X by the list returned by f:

```
>> map(X, f);

   [[x1 y1, x1 y2, x1 y3], [x2 y1, x2 y2, x2 y3],

    [y1 x3, x3 y2, x3 y3]]
```

Exercise 4.17: For each m, we use the sequence generator $ to create a list of all integers to be checked. Then we extract all primes from the list via `select(.,isprime)`. The number of primes is just `nops` of the resulting list. We compute this value for all m between 0 and 41:

```
>> nops(select([(n^2 + n + m) $ n = 1..100], isprime))
    $ m = 0..41;
```

```
   1, 32, 0, 14, 0, 29, 0, 31, 0, 13, 0, 48, 0, 18, 0, 11,

      0, 59, 0, 25, 0, 14, 0, 28, 0, 28, 0, 16, 0, 34,

      0, 35, 0, 11, 0, 24, 0, 36, 0, 17, 0, 86
```

There is a simple explanation for the zero values for even $m > 0$. Since $n^2 + n = n(n+1)$ is always even, $n^2 + n + m$ is an even integer greater than 2 and hence not a prime.

Exercise 4.18: We store the children in a list C and remove the one that has been counted out at the end of each round. For technical reasons, we represent the positions $1, 2, \ldots, n$ in a list of n children by the integers $0, 1, \ldots, n - 1$. Let out $\in \{0, 1, \ldots, n - 1\}$ be the position of the last child that was counted out. In the next round, we are at position out+m-1 in the current list (which is shorter by 1 than the previous one) after m words. Since we are counting cyclically, we take this value modulo the number of remaining children:

```
>> m := 9: n := 12: C := [$ 1..n]: out := 0:
>> out := (out + m - 1) mod nops(C):
>> C[out + 1]; unassign(C[out + 1]):
```

```
   9
```

```
>> out := (out + m - 1) mod nops(C):
>> C[out + 1]; unassign(C[out + 1]):
```

```
   6
```

It is useful to implement this in form of a loop (Chapter 16):

```
>> m := 9: n := 12: C := [$ 1..n]: out := 0:
```

```
>> repeat
     out := (out + m - 1) mod nops(C):
     print(C[out + 1]):
     unassign(C[out + 1])
   until nops(C) = 0 end_repeat:
```

9

6

...

1

2

Exercise 4.19: The following two conversions *list* ↦ *set* ↦ *list* in general change the order of the list elements:

```
>> set := {op(list)}:  list := [op(set)]:
```

Exercise 4.20:

```
>> A := {a, b, c}: B := {b, c, d}: C := {b, c, e}:
>> A union B union C, A intersect B intersect C,
   A minus (B union C);
```

{a, b, c, d, e}, {b, c}, {a}

Exercise 4.21: You obtain the union via _union:

```
>> M := {{2, 3}, {3, 4}, {3, 7}, {5, 3}, {1, 2, 3, 4}}:
>> _union(op(M));
```

{1, 2, 3, 4, 5, 7}

and the intersection via _intersect:

```
>> _intersect(op(M));
```

> {3}

Exercise 4.22: The function `combinat::choose(M, k)` returns a sequence of all subsets with k elements of a set M:

```
>> M := {i $ i = 5..20}:
>> subsets := combinat::choose(M, 3):
```

The number of such subsets is:

```
>> nops(subsets);
```

> 560

Since the number of subsets with n elements of a set with m elements is the binomial coefficient $\binom{m}{n}$, you need not generate the subsets to count them:

```
>> binomial(nops(M), 3);
```

> 560

Exercise 4.23:

```
>> telephoneDirectory := table(Ford = 1815,
     Reagan = 4711, Bush = 1234, Clinton = 5678):
```

An indexed call returns Ford's number:

```
>> telephoneDirectory[Ford];
```

> 1815

You can extract all table entries containing the number 5678 via `select`:

```
>> select(telephoneDirectory, has, 5678);

   table(
     Clinton = 5678
   )
```

Exercise 4.24: The command [op(Table)] returns a list of all assignment equations. The call map(.,op,i) $(i = 1, 2)$ extracts the left and the right hand sides, respectively, of the equations:

```
>> T := table(a = 1, b = 2,
                1 - sin(x) = "derivative of x + cos(x)" ):
>> indices := map([op(T)], op, 1);

   [a, b, 1 - sin(x)]

>> values := map([op(T)], op, 2);

   [1, 2, "derivative of x + cos(x)"]
```

Exercise 4.25: The following timings (in milliseconds) show that generating a table is more time consuming:

```
>> n := 100000:
>> time( (T := table( (i=i) $ i=1..n)) ),
   time( (L := [i $ i=1..n]) );

   3750, 820
```

However, working with tables is notably faster. The following assignments create an additional table entry and extend the list by one element, respectively:

```
>> time( (T[n + 1] := New) ), time( (L := L.[New]) );

   10, 140
```

Exercise 4.26: We use the sequence generator $ to create a nested list and pass it to **array**:

```
>> n := 20: array(1..n, 1..n,
                  [[1/(i + j -1) $ j = 1..n] $ i = 1..n]):
```

Exercise 4.27:

```
>> TRUE and (FALSE or not (FALSE or not FALSE));

   FALSE
```

Exercise 4.28: We use the function **zip** to generate a list of comparisons. We pass the system function **_less** as third argument, which generates inequalities of the form a<b. Then we extract the sequence of inequalities via **op** and pass it to **_and**:

```
>> L1 := [ 10*i^2 - i^3 $ i = 1..10]:
>> L2 := [ i^3 + 13*i $ i = 1..10]:
>> _and(op(zip(L1, L2, _less)));

   9 < 14 and 32 < 34 and 63 < 66 and 96 < 116 and

      125 < 190 and 144 < 294 and 147 < 434 and 128 < 616

      and 81 < 846 and 0 < 1130
```

Finally, evaluating this expression by means of **bool** answers the question:

```
>> bool(%);

   TRUE
```

Exercise 4.29: The function `sort` does not sort the identifiers alphabetically by their names but according to an internal order (Section 4.6). Thus we convert them to strings via `expr2text` before sorting:

```
>> [op(anames(3))]: map(%, expr2text): sort(%);

   ["Ax", "Axiom", "AxiomConstructor", "BackSubstitution",

    "Cat", "Category", ... , "zeta", "zip"]
```

Exercise 4.30: We compute the reflection of the palindrome

```
>> text := "Never odd or even":
```

by passing the reflected sequence of individual characters to the function `_concat`, which converts it to a string again:

```
>> n := strlen(text): _concat(text[n - i] $ i = 1..n);

   "neve ro ddo reveN''
```

This can also be achieved with the call `revert(text)`.

Exercise 4.31:

```
>> f := x -> (x^2): g := x -> (sqrt(x)):
>> (f@f@g)(2), (f@@100)(x);

      1267650600228229401496703205376
   4, x
```

Exercise 4.32: If you use the arrow operator to define a function in MuPAD version 1.4, then you should use `hold` to prevent the immediate evaluation of system functions:

```
>> f := L -> hold([L[nops(L) + 1 - i] $ i = 1..nops(L)]);

   L -> [(nops(L) + 1) - i] $ i = 1..nops(L)]
```

```
>> f([a, b, c]);

   [c, b, a]
```

In versions of MuPAD beyond 1.4, the definition

```
>> f := L -> [L[nops(L)+1-i] $ i = 1..nops(L)]:
```

yields the desired function. However, the simplest solution is to use
f:=revert.

Exercise 4.33: You can use the function last (Chapter 12) to generate the Chebyshev polynomials as expressions:

```
>> T0 := 1: T1 := x:
>> T2 := 2*x*% - %2; T3:= 2*x*% - %2; T4:= 2*x*% - %2;
```

$$2 x^2 - 1$$

$$2 x (2 x^2 - 1) - x$$

$$2 x (2 x (2 x^2 - 1) - x) - 2 x^2 + 1$$

A much more elegant way is to translate the recursive definition into a recursive function:

```
>> T := (k, x) ->
       (if k < 2
           then x^k
           else 2*x*T(k - 1, x) - T(k - 2, x)
       end_if):
```

Then we obtain:

```
>> T(i, 1/3) $ i = 2..5;

   -7/9, -23/27, 17/81, 241/243
```

```
>> T(i, 0.33) $ i = 2..5;

   -0.7822, -0.846252, 0.22367368, 0.9938766288

>> T(i, x) $ i = 2..5;

      2            2
   2 x  - 1, 2 x (2 x  - 1) - x,


            2                  2
   2 x (2 x (2 x  - 1) - x) - 2 x  + 1,


             2                           2
   x - 2 x (2 x  - 1) + 2 x (2 x (2 x (2 x  - 1) - x) -


      2
   2 x  + 1)
```

You can obtain expanded representations of the polynomials by inserting a call to **expand** (Section 9.1) in the function definition. The Chebyshev polynomials are already implemented in the library **orthpoly** for orthogonal polynomials. Then the call **orthpoly::chebyshev1(i,x)** returns the ith Chebyshev polynomial.

Exercise 4.34: In principle, you can compute the derivatives of f in MuPAD and substitute $x = 0$. However, it is simpler to approximate the function by a Taylor series whose leading terms describe the behavior in the neighborhood of $x = 0$:

```
>> taylor(tan(sin(x)) - sin(tan(x)), x = 0, 10);

    7        9
   x     29 x          10
   -- + ----- + O(x  )
   30    756
```

Thus, $f(x) = x^7/30 \cdot (1 + O(x^2))$, and hence f has a root of order 7 at $x = 0$.

Exercise 4.35: The reason for the difference between the following two results:

```
>> taylor(cos(x), x), diff(taylor(sin(x), x), x);
```

$$1 - \frac{x^2}{2} + \frac{x^4}{24} + O(x^6), \quad 1 - \frac{x^2}{2} + \frac{x^4}{24} + O(x^5)$$

is the environment variable ORDER with the default value 6. The first call computes the cosine series up to $O(x^6)$, and it turns out that there is no term of order x^5. The call

```
>> taylor(sin(x), x);
```

$$x - \frac{x^3}{6} + \frac{x^5}{120} + O(x^6)$$

does not realize that the sine series has no term of order x^6 and that $O(x^7)$ would be correct as well. The $O(x^5)$ appears when the (not really existing) term $O(x^6)$ is differentiated.

Exercise 4.36: An asymptotic expansion yields:

```
>> f := sqrt(x + 1) - sqrt(x - 1):
>> g := series(f, x = infinity, 10);
```

$$\left(\frac{1}{x}\right)^{1/2} + \frac{\left(\frac{1}{x}\right)^{5/2}}{8} + \frac{7\left(\frac{1}{x}\right)^{9/2}}{128} + \frac{33\left(\frac{1}{x}\right)^{13/2}}{1024} +$$

$$O\left(\left(\frac{1}{x}\right)^{17/2}\right)$$

Thus

$$f \approx \frac{1}{\sqrt{x}} \left(1 + \frac{1}{8\,x^2} + \frac{7}{128\,x^4} + \frac{33}{1024\,x^6} + \cdots \right),$$

and hence $f(x) \approx 1/\sqrt{x}$ for all real $x \gg 1$. The next better approximation is $f(x) \approx \dfrac{1}{\sqrt{x}} \left(1 + \dfrac{1}{8\,x^2} \right)$.

Exercise 4.37: The command ?revert requests the corresponding help page.

```
>> f := taylor(sin(x + x^3), x); g := revert(%);
```

```
        3        5
     5 x     59 x         6
 x + ---- - ------ + O(x )
      6      120
```

```
        3        5
     5 x     103 x        6
 x - ---- + ------ + O(x )
      6       40
```

To check this result, we consider the composition of f and g, whose series expansion is that of the identity function $x \mapsto x$:

```
>> g@f;
```

```
        6
 x + O(x)
```

Exercise 4.38: We perform the computation over the standard coefficient ring (Section 4.15.1), which comprises both rational numbers and floating-point numbers:

```
>> n := 15:
>> H := Dom::Matrix()(n, n, (i, j) -> ((i + j -1)^(-1))):
>> e := Dom::Matrix()(n, 1, 1): b := H*e:
```

We first compute the solution of the system of equations $H\mathbf{x} = \mathbf{b}$ with exact arithmetic over the rational numbers. Then we convert all entries of H and b to floating-point numbers and solve the system numerically:

```
>> exact = H^(-1)*b, numerical = float(H)^(-1)*float(b);
```

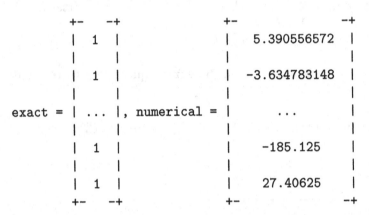

The errors in the numerical solution originate from rounding errors, which are version dependent. To demonstrate this, we repeat the numerical computation with higher precision. Since parts of the inversion algorithm are implemented with **option remember** (Section 18.8), a new computation without restart would use the values that were computed with a lower precision. For this reason, we restart the MuPAD session via **reset()** (Section 14.3):

```
>> reset(): DIGITS := 20: n :=15:
>> H := Dom::Matrix()(n, n, (i, j) -> ((i + j -1)^(-1))):
>> e := Dom::Matrix()(n, 1, 1): b := H*e:
>> numerical = float(H)^(-1)*float(b);
```

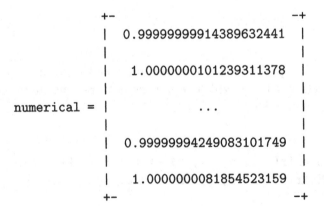

Exercise 4.39: We look for values where the determinant of the matrix vanishes:

```
>> Dom::Matrix()([[1, a, b], [1, 1, c ], [1, 1, 1]]):
>> Factor(linalg::det(%));
```

$$(a - 1)\ (c - 1)$$

Thus the matrix is invertible unless $a = 1$ or $c = 1$.

Exercise 4.40: We first store the matrix data in arrays. These arrays are used later to generate matrices over different coefficient rings:

```
>> a := array(1..3, 1..3, [[ 1, 3, 0],
                           [-1, 2, 7],
                           [ 0, 8, 1]]):
>> b := array(1..3, 1..2, [[7, -1], [2, 3], [0, 1]]):
```

To simplify the following function calls we export the library Dom:

```
>> export(Dom):
```

Now we define the constructor MQ for matrices over the rational numbers and convert the arrays to corresponding matrices:

```
>> MQ := Matrix(Rational): A := MQ(a): B := MQ(b):
```

The method **transpose** of the constructor determines the transpose of a matrix B via MQ::transpose(B):

```
>> (2*A + B*MQ::transpose(B))^(-1);
```

```
+-                                      -+
|   34/1885,  7/1508,   -153/7540  |
|                                       |
|   11/3770, -31/3016,  893/15080  |
|                                       |
|  -47/3770, 201/3016, -731/15080  |
+-                                      -+
```

Computing over the residue class ring modulo 7 we find:

```
>> Mmod7 := Matrix(IntegerMod(7)):
>> A := Mmod7(a): B := Mmod7(b):
>> C := (2*A + B*Mmod7::transpose(B)): C^(-1);
```

```
    +-                              -+
    |  3 mod 7, 0 mod 7, 1 mod 7  |
    |                              |
    |  1 mod 7, 3 mod 7, 2 mod 7  |
    |                              |
    |  4 mod 7, 2 mod 7, 2 mod 7  |
    +-                              -+
```

We check this by multiplying the inverse by the original matrix, which
yields the identity matrix over the coefficient ring:

```
>> %*C;
```

```
    +-                              -+
    |  1 mod 7, 0 mod 7, 0 mod 7  |
    |                              |
    |  0 mod 7, 1 mod 7, 0 mod 7  |
    |                              |
    |  0 mod 7, 0 mod 7, 1 mod 7  |
    +-                              -+
```

Exercise 4.41: We compute over the coefficient ring of rational numbers:

```
>> MQ := Dom::Matrix(Dom::Rational):
```

To define the matrix we pass a function associating indices with matrix
entries to the constructor:

```
>> A := MQ(3, 3, (i, j) -> (if i=j then 0 else 1 end_if));
```

```
+-          -+
|  0, 1, 1  |
|           |
|  1, 0, 1  |
|           |
|  1, 1, 0  |
+-          -+
```

The determinant of A is

```
>> linalg::det(A);
```

   2

The eigenvalues are the roots of the characteristic polynomial p (the MuPAD data structure poly for polynomials is discussed in Section 4.16)[1]:

```
>> p := linalg::charPolynomial(A, x);
```

```
         3
   poly(x  + (-3) x - 2, [x])
```

```
>> solve(p = 0, x);
```

   {-1, 2}

Alternatively, the linalg package provides a function for computing eigenvalues:

```
>> linalg::eigenValues(A);
```

   {-1, 2}

Let E denote the 3×3 identity matrix. The eigenspace for the eigenvalue $\lambda \in \{-1, 2\}$ is the solution space of the system of linear equations $(A - \lambda \cdot E)\, \mathbf{x} = \mathbf{0}$. The solution vectors span the nullspace (the "kernel")

---

[1] In MuPAD versions beyond 1.4, the function linalg::charPolynomial is renamed to linalg::charpoly. Moreover, it returns always a polynomial of domain type Dom::DistributedPolynomial.

of the matrix $A - \lambda \cdot E$. The function linalg::nullSpace[2] computes a basis for the kernel of a matrix:

```
>> Id := MQ(3, 3, 1, Diagonal):
>> lambda := -1: linalg::nullSpace(A - lambda*Id);
```

```
-- +-      -+ +-      -+ --
|  |  -1  |  |  -1  |  |
|  |      |  |      |  |
|  |   1  |, |   0  |  |
|  |      |  |      |  |
|  |   0  |  |   1  |  |
-- +-      -+ +-      -+ --
```

There are two linearly independent basis vectors. Hence the eigenspace for the eigenvalue $\lambda = -1$ is two-dimensional. The other eigenvalue is simple:

```
>> lambda := 2: linalg::nullSpace(A - lambda*Id);
```

```
-- +-      -+ --
|  |   1  |  |
|  |      |  |
|  |   1  |  |
|  |      |  |
|  |   1  |  |
-- +-      -+ --
```

Alternatively linalg::eigenVectors computes all eigenspaces simultaneously:

```
>> linalg::eigenVectors(A);
```

---

[2] In MuPAD versions beyond 1.4, the three functions linalg::eigenValues, linalg::eigenVectors, and linalg::nullSpace are renamed to linalg::eigenvalues, linalg::eigenvectors, and linalg::nullspace, respectively.

```
-- --           -- +-   -+ -- --
|  |            |  |  1 |  |  |  |
|  |            |  |    |  |  |  |
|  |  2, 1,  |  |  1 |  |  |  |,
|  |            |  |    |  |  |  |
|  |            |  |  1 |  |  |  |
-- --           -- +-   -+ -- --
```

```
--              -- +-   -+ +-   -+ -- -- --
|               |  | -1 |  | -1 |  |  |  |
|               |  |    |  |    |  |  |  |
|  -1, 2,  |  |  1 |, |  0 |  |  |  |
|               |  |    |  |    |  |  |  |
|               |  | 0 |  |  1 |  |  |  |
--              -- +-   -+ +-   -+ -- -- --
```

The return value is a nested list. For each eigenvalue λ, it contains a list of the form

$$[\,\lambda\,,\text{ multiplicity of }\lambda\,,\text{ eigenspace basis}\,].$$

Exercise 4.42:

```
>> p := poly(x^7 - x^4 + x^3 - 1): q := poly(x^3 - 1):
>> p - q^2;

      7           6           4        3
  poly(x  + (-1) x  + (-1) x  + 3 x  - 2, [x])
```

The polynomial p is a multiple of q:

```
>> p/q;

       4
  poly(x  + 1, [x])
```

This is confirmed by a factorization:

```
>> Factor(p);
```

$$
\text{poly}(x - 1, \ [x]) \ \text{poly}(x \overset{2}{+} x + 1, \ [x])
$$

$$
\text{poly}(x \overset{4}{+} 1, \ [x])
$$

```
>> Factor(q);
```

$$
\text{poly}(x - 1, \ [x]) \ \text{poly}(x \overset{2}{+} x + 1, \ [x])
$$

Exercise 4.43: We only need to try the possible remainders $0, 1, 2$ modulo 3 for the coefficients a, b, c in $a x^2 + b x + c$. We generate a list of all 18 quadratic polynomials with $a \neq 0$ as follows:

```
>> p := 3: K := Dom::IntegerMod(p):
>> [((poly(a*x^2 + b*x + c, [x], K) $ a = 1..p-1
   ) $ b = 0..p-1) $ c = 0..p-1]:
```

The command select(.,irreducible) extracts the 6 irreducible polynomials:

```
>> select(%, irreducible);
```

$$
[\text{poly}(x \overset{2}{+} 1, \ [x], \ \text{Dom::IntegerMod}(3)),
$$

$$
\text{poly}(2 x \overset{2}{+} x + 1, \ [x], \ \text{Dom::IntegerMod}(3)),
$$

$$
\text{poly}(2 x \overset{2}{+} 2 x + 1, \ [x], \ \text{Dom::IntegerMod}(3)),
$$

$$
\text{poly}(2 x \overset{2}{+} 2, \ [x], \ \text{Dom::IntegerMod}(3)),
$$

$$\text{poly}(x^2 + x + 2, [x], \text{Dom}::\text{IntegerMod}(3)),$$

$$\text{poly}(x^2 + 2x + 2, [x], \text{Dom}::\text{IntegerMod}(3))]$$

Exercise 5.1: The value of x is the identifier a1. The evaluation of x yields the identifier c1. The value of y is the identifier b2. The evaluation of y yields the identifier c2. The value of z is the identifier a3. The evaluation of z yields 10.

The evaluation of u1 leads to an infinite recursion, which MuPAD aborts with an error message. The evaluation of u2 yields the expression v2^2 - 1.

Exercise 6.1: The result of subsop(b+a,1=c) is b+c and not c+a, as you might have expected. The reason is that subsop evaluates its arguments. The system reorders the sum internally when evaluating it, and thus subsop processes a+b instead of b+a. Upon return, the result c+b is reordered again.

Exercise 6.2: The highest derivative occurring in g is the 6th derivative diff(f(x),x$6). We pass the sequence of replacement equations:

```
diff(f(x), x $ 6) = f6, diff(f(x), x $ 5) = f5, ... ,
                    f(x) = f0
```

to MuPAD's substitution function. Note that according to the usual mathematical notation, diff returns the function itself as the 0th derivative: diff(f(x),x$0)=diff(f(x))=f(x).

```
>> unassign(f): g := diff(f(x)/diff(f(x),x), x $ 5):
>> subs(g, (diff(f(x), x $ 6-i) = f.(6-i)) $ i = 0..6);
```

$$\frac{25\ f2\ f4}{f1^2} - \frac{4\ f5}{f1} - \frac{f0\ f6}{f1^2} + \frac{10\ f0\ f2\ f5}{f1^3} + \frac{20\ f0\ f3\ f4}{f1^3} +$$

$$\frac{20\ f3^2}{f1^2} + \frac{60\ f2^4}{f1^4} - \frac{100\ f2^2\ f3}{f1^3} - \frac{120\ f0\ f2^5}{f1^6} -$$

$$\frac{90\ f0\ f2^2\ f3}{f1^4} - \frac{60\ f0\ f2^2\ f4}{f1^4} + \frac{240\ f0\ f2^3\ f3}{f1^5}$$

Exercise 7.1: The following commands yield the desired evaluation of the function:

```
>> f := sin(x)/x: x := 1.23: f;
```

    0.7662510584

However, x now has a value. The following call `diff(f,x)` would internally lead to the invalid command `diff(0.7662510584,1.23)`, since `diff` evaluates its arguments. You can circumvent this problem by preventing a complete evaluation of the arguments via **level** or **hold** (Section 5.2):

```
>> g := diff(level(f,1), hold(x)); g;
```

$$\frac{\cos(x)}{x} - \frac{\sin(x)}{x^2}$$

    -0.3512303507

Here the evaluation of `hold(x)` is the identifier x and not its value. Writing `hold(f)` instead of `level(f,1)` would yield the wrong result `diff(hold(f),hold(x))=0` since `hold(f)` does not contain `hold(x)`. The expression `level(f,1)` replaces f by its value `sin(x)/x` (Section 5.2). The next call of g returns the evaluation of g, namely the value of the derivative at x=1.23. Alternatively you can delete the value of x:

```
>> unassign(x): diff(f,x): subs(%, x = 1.23); eval(%);

   0.81300813 cos(1.23) - 0.6609822195 sin(1.23)

   -0.3512303507
```

Exercise 7.2: The first three derivatives of the numerator and the denominator vanish at the point $x = 0$:

```
>> Z := x -> (x^3*sin(x)): N := x -> ((1 - cos(x))^2):
>> Z(0), N(0), Z'(0), N'(0), Z''(0), N''(0),
   Z'''(0), N'''(0);

   0, 0, 0, 0, 0, 0, 0, 0
```

For the fourth derivatives, we have:

```
>> Z''''(0), N''''(0);

   24, 6
```

Thus the limit is $Z''''(0)/N''''(0) = 4$, according to de l'Hospital's rule. The function limit computes the same result:

```
>> limit(Z(x)/N(x), x = 0);

   4
```

Exercise 7.3: The first order partial derivatives of f_1 are:

```
>> f1 := sin(x1*x2): diff(f1, x1), diff(f1, x2);

   x2 cos(x1 x2), x1 cos(x1 x2)
```

The second order derivatives are:

```
>> diff(f1, x1, x1), diff(f1, x1, x2),
   diff(f1, x2, x1), diff(f1, x2, x2);
```

$$- x2^2 \; \sin(x1 \; x2), \; \cos(x1 \; x2) - x1 \; x2 \; \sin(x1 \; x2),$$

$$\cos(x1 \; x2) - x1 \; x2 \; \sin(x1 \; x2), \; - x1^2 \; \sin(x1 \; x2)$$

The total derivative of f_2 with respect to t is:

```
>> f2 := x^2*y^2: x := sin(t): y := cos(t): diff(f2, t);
```

$$2 \; \cos(t)^3 \; \sin(t) - 2 \; \cos(t) \; \sin(t)^3$$

Exercise 7.4:

```
>> int(sin(x)*cos(x), x = 0..PI/2),
   int(1/(sqrt(1 - x^2)), x = 0..1),
   int(x*atan(x), x = 0..1);
```

$$1/2, \; \frac{PI}{2}, \; \frac{PI}{4} - 1/2$$

The function `atan` is called `arctan` in MuPAD versions beyond 1.4. In MuPAD 1.4.2, the system returns a complex representation for the following integral:

```
>> s := int(1/x, x = -2..-1);
```

$$I \; PI - \ln(-2)$$

The function `rectform` (Section 9.1) decomposes the result into its real and imaginary parts, and we see that it is in fact real:

```
>> rectform(s);
```

$$-\ln(2)$$

Exercise 7.5:

```
>> int(x/(2*a*x - x^2)^(3/2), x);

        2
    x (x  - 2 a x)
  - -----------------
              2 3/2
    a (2 a x - x )

>> int(sqrt(x^2 - a^2), x);

      2    2 1/2    2          2    2 1/2
    x (x  - a )     a  ln(x + (x  - a )  )
    -------------- - -----------------------
         2                     2

>> int(1/(x*sqrt(1 + x^2)), x);

          /      1      \
  - atanh| ----------- |
         |    2    1/2 |
         \ (x  + 1)    /
```

Exercise 7.6: The function `changevar` only performs a change of variables, without invoking the integration:

```
>> changevar(int(sin(x)*sqrt(1 + sin(x)), x=-PI/2..PI/2),
             sin(x) = t);

     /            1/2           \
     | t (t + 1)                |
  int| ------------, t = -1..1 |
     |       2 1/2              |
     \ (1 - t )                 /
```

Only another evaluation activates the integration routine:

```
>> eval(%): % = float(%);
```

```
   1/2
 2 2
 ------ =   0.9428090415
   3
```

Numerical quadrature returns the same result:

```
>> numeric::quadrature(sin(x)*sqrt(1 + sin(x)),
                  x = -PI/2..PI/2);


   0.9428090415
```

Exercise 8.1: The equation solver returns the general solution:

```
>> equations := {a +   b +   c +   d +   e = 1,
                 a + 2*b + 3*c + 4*d + 5*e = 2,
                 a - 2*b - 3*c - 4*d - 5*e = 2,
                 a -   b -   c -   d -   e = 3}:
>> solve(equations, {a, b, c, d, e});

   {{a = 2, b = d + 2 e - 3, c = 2 - 2 d - 3 e}}
```

The free parameters are on the right hand sides of the solved equations. You can determine them in MuPAD by extracting the right hand sides and using **indets** to find the identifiers contained therein:

```
>> map(%, map, op, 2); indets(%);

   {{d + 2 e - 3, 2 - 2 d - 3 e, 2}}

   {d, e}
```

Exercise 8.2: In MuPAD 1.4.2, the symbolic solution is:

```
>> solution := solve(ode(
     {y'(x)=y(x) + z(x), z'(x) = y(x)}, {y(x), z(x)}));
```

```
{                    /          1/2 \     /          1/2 \
{                 | x     x 5    |       | x     x 5    |
{           C1 exp| - -  ------  |  C2 exp| - +  ------ |
{                 \ 2      2    /         \ 2      2   /
{ y(x) = ----------------------------  +  -------------------------- -
{                     2                            2
```

```
          /          1/2 \              /          1/2 \
   1/2    | x     x 5    |       1/2    | x     x 5    |
 C1 5  exp| - -  ------  |     C2 5  exp| - +  ------ |
          \ 2      2    /               \ 2      2   /
 -----------------------------  +  --------------------------- ,
              2                              2
```

```
                                                          }
                                                          }
          /          1/2 \              /          1/2 \ }
          | x     x 5    |              | x     x 5    | }
 z(x) = C1 exp| - -  ------  |  + C2 exp| - +  ------ | }
          \ 2      2    /               \ 2      2   / }
```

with free constants C1, C2. Now we set $x = 0$ and substitute y(0)
and z(0), respectively, for the initial conditions. Then we solve the
resulting linear system of equations for C1 and C2:

```
>> solve(eval(subs(solution, x=0, y(0)=1, z(0)=1),
        {C1, C2}));
```

```
{ {                 1/2        1/2        } }
{ {                5          5           } }
{ { C1 = 1/2 -  ----,  C2 =  ---- + 1/2 } }
{ {                10         10          } }
```

We remove the outer curly braces via op, and assign the solution values
to C1 and C2 by means of assign:

```
>> assign(op(%)):
```

Thus the value at $x = 1$ of the symbolic solution for the above initial
conditions is:

```
>> x := 1:
>> [normal(op(solution, 1)), normal(op(solution, 2))];

    --            /       1/2 \      /  1/2          \
    |             |      5    |      | 5             |
    |         exp| 1/2 - ---- |   exp| ---- + 1/2 |
    |             \       2   /      \  2           /
    | y(1) = ------------------- + ------------------ -
    --                2                     2


             /       1/2 \              /  1/2          \
     1/2     |      5    |       1/2    | 5             |
    3 5   exp| 1/2 - ---- |     3 5   exp| ---- + 1/2 |
             \       2   /              \  2           /
    ------------------------- + -------------------------,
               10                          10


             /       1/2 \      /  1/2          \
             |      5    |      | 5             |
         exp| 1/2 - ---- |   exp| ---- + 1/2 |
             \       2   /      \  2           /
    z(1) = ------------------- + ------------------ -
                   2                     2


             /       1/2 \              /  1/2          \ --
     1/2     |      5    |       1/2    | 5             | |
    5     exp| 1/2 - ---- |     5     exp| ---- + 1/2 | |
             \       2   /              \  2           / |
    ------------------------- + ------------------------- |
               10                          10            --
```

Finally, we apply **float** to the right hand sides of these equations:

```
>> map(%, float@op, 2);

    [5.812568463, 3.798245729]
```

Exercise 8.3:

1)

```
>> solve(ode(y'(x)/y(x)^2 = 1/x, y(x)));

    {       1       }
    { ------------- }
    { - C1 - ln(x)  }
```

2a)

```
>> solve(ode({y'(x) - sin(x)*y(x) = 0, D(y)(1)=1}, y(x)));

    {      exp(-cos(x))      }
    { --------------------- }
    { sin(1) exp(-cos(1))   }
```

2b)

```
>> solve(ode({2*y'(x) + y(x)/x = 0, D(y)(1) = PI}, y(x)));

    {   2 PI }
    { - ---- }
    {   1/2  }
    {   x    }
```

3)

```
>> solve(ode({diff(x(t),t) = -3*y(t)*z(t),
              diff(y(t),t) =  3*x(t)*z(t),
              diff(z(t),t) = -x(t)*y(t)},
             {x(t),y(t),z(t)}));
```

$$
\{[x(t) = (3\ z(t)^2 - C4)^{1/2}, \ y(t) = (-\ C5 - 3\ z(t)^2)^{1/2}
$$

$$
], \ [x(t) = -\ (3\ z(t)^2 - C4)^{1/2},
$$

```
                       2 1/2
    y(t) = (- C5 - 3 z(t) )    ], [


              2       1/2
    x(t) = (3 z(t)  - C4)    ,


                         2 1/2
    y(t) = - (- C5 - 3 z(t) )    ], [


              2       1/2
    x(t) = - (3 z(t)  - C4)    ,


                         2 1/2
    y(t) = - (- C5 - 3 z(t) )    ]}
```

Exercise 8.4: The function `solve` directly yields the solution of the recurrence:

```
>> solve(rec(F(n) = F(n-1) + F(n-2), F(n),
            {F(0) = 0, F(1) = 1}));

   {        /  1/2      \n            /           1/2 \n }
   {  1/2  | 5          |       1/2  |         5    |  }
   { 5     | ---- + 1/2 |    5       | 1/2 -  ---- |  }
   {       \  2         /            \          2  /  }
   { -------------------- - -------------------- }
   {          5                      5             }
```

Exercise 9.1: You obtain the answer immediately from:

```
>> simplify(cos(x)^2 + sin(x)*cos(x));

  cos(2 x)    sin(2 x)
  -------- + -------- + 1/2
     2          2
```

You get the same result by applying **combine** to rewrite products of trigonometric functions as sums:

```
>> combine(cos(x)^2 + sin(x)*cos(x), sincos);
```

$$\frac{\cos(2\ x)}{2} + \frac{\sin(2\ x)}{2} + 1/2$$

Exercise 9.2:
1)

```
>> expand(cos(5*x)/(sin(2*x)*cos(x)^2));
```

$$\frac{\cos(x)^2}{2\ \sin(x)} - 5\ \sin(x) + \frac{5\ \sin(x)^3}{2\ \cos(x)^2}$$

2)

```
>> f := (sin(x)^2 - exp(2*x)) /
      (sin(x)^2 + 2*sin(x)*exp(x) + exp(2*x)):
>> normal(expand(f));
```

$$\frac{\sin(x) - \exp(x)}{\sin(x) + \exp(x)}$$

3)

```
>> f := (sin(2*x) - 5*sin(x)*cos(x)) /
      (sin(x)*(1 + tan(x)^2)):
>> combine(normal(expand(f)), sincos);
```

$$-\frac{9\ \cos(x)}{4} - \frac{3\ \cos(3\ x)}{4}$$

4)

```
>> f := sqrt(14 + 3*sqrt(3 +
                    2*sqrt(5 - 12*sqrt(3 - 2*sqrt(2))))):
>> simplify(f, sqrt);
```

$$2^{1/2} + 3$$

Exercise 9.3: As a first step we perform a normalization:

```
>> int(sqrt(sin(x) + 1), x): normal(diff(%, x));
```

$$\frac{\cos(x)^2 - 2\sin(x) + 2\sin(x)^3 + 3\cos(x)^2\sin(x)}{\cos(x)^2(\sin(x) + 1)^{1/2}}$$

Then we eliminate the cosine terms:

```
>> subs(%, cos(x)^2 = 1 - sin(x)^2);
```

$$(2\sin(x)^3 - 2\sin(x) - \sin(x)^2 + 3\sin(x)(1 - \sin(x)^2) + 1) / (\cos(x)^2(\sin(x) + 1)^{1/2})$$

The system does not replace the expression in the denominator since it is contained as $\cos(x)^{(-2)}$ in the expression tree and not as $\cos(x)^2$:

```
>> subs(%, cos(x)^(-2) = (1 - sin(x)^2)^(-1));
```

$$(2 \sin(x)^3 - 2 \sin(x) - \sin(x)^2 +$$

$$3 \sin(x) (1 - \sin(x))^2 + 1) /$$

$$((\sin(x) + 1)^{1/2} (1 - \sin(x))^2))$$

The final normalization step achieves the desired simplification:

```
>> normal(%);
```

$$(\sin(x) + 1)^{1/2}$$

Exercise 9.4: The function `assume` (Section 9.3) assigns properties to identifiers. These are taken into account by `limit`:

```
>> assume(a > 0): limit(x^a, x = infinity);

   infinity

>> limit(x^0, x = infinity);

   1

>> assume(a < 0): limit(x^a, x = infinity);

   0
```

Exercise 10.1: In analogy to the previous gcd example, we obtain the following experiment:

```
>> die := random(1..6):
>> experiment := [[die(), die(), die()] $ i = 1..216]:
>> diceScores := map(experiment, x -> (x[1]+x[2]+x[3])):
>> frequencies := Dom::Multiset(op(diceScores)):
```

```
>> sortingOrder := (x, y) -> (x[1] < y[1]):
>> sort([op(frequencies)], sortingOrder);
```

[[4, 4], [5, 9], [6, 8], [7, 9], [8, 16], [9, 20],

[10, 27], [11, 31], [12, 32], [13, 20], [14, 13],

[15, 12], [16, 6], [17, 7], [18, 2]]

In this experiment the score 3 did not occur.

Exercise 10.2: a) The command

```
>> r := float@random(0..10^10)/10^10:
```

creates a generator for random numbers in $[0, 1]$. Thus

```
>> n := 1000: absValues := [sqrt(r()^2+r()^2) $ i = 1..n]:
```

returns a list with the absolute values of n random vectors in the rectangle $Q = [0, 1] \times [0, 1]$. The number of values ≤ 1 is the number of random points in the right upper quadrant of the unit circle:

```
>> m := nops(select(absValues, z -> (z <= 1)));
```

787

Since m/n approximates the area $\pi/4$ of the right upper quadrant of the unit circle, we obtain the following approximation to π:

```
>> float(4*m/n);
```

3.148

b) First we determine the maximum of f. The following function plot shows that f is monotonically increasing on the interval $[0, 1]$:

```
>> f := x*sin(x) + cos(x)*exp(x): plotfunc(f, x=0..1):
```

Thus f assumes its maximal value at the right end of the interval. Therefore $M = f(1)$ is an upper bound for the function:

```
>> M := float(subs(f, x = 1));
```

   2.310164924

We use the random number generator defined above to generate random points in the rectangle $[0, 1] \times [0, M]$:

```
>> n := 1000: pointlist := [[r(), M*r()] $ i = 1..n]:
```

We select those points $p = [x, y]$ for which $0 \leq y \leq f(x)$ holds:

```
>> select(pointlist, p -> (p[2] <= subs(f, x = p[1]))):
>> m := nops(%);
```

   740

Thus the following is an approximation of the integral:

```
>> m/n*M;
```

   1.709522044

The exact value is:

```
>> float(int(f, x = 0..1));
```

   1.679193292

Exercise 11.1: The MuPAD function **round** rounds a real number to the nearest integer. The default value in **plotfunc** for the number of graphical sample points is too small to cover the discontinuities of f exactly:

```
>> f := abs(x - round(x))/x: plotfunc(f, x = 1..30);
```

If you use **plot2d** then you can increase the resolution with the option **Grid=[..]**:

```
>> plot2d(Scaling = UnConstrained,
          [Mode = Curve, [x, f], x = [1, 30],
          Grid = [1000]]);
```

Exercise 11.2: We supply `Scaling=Constrained` as a scene option to prevent the sphere from looking like an ellipsiod:

```
>> sphere := [Mode=Surface,
              [cos(u)*sin(v), sin(u)*sin(v), cos(v)],
              u=[0,2*PI], v=[0,PI],
              Style = [ColorPatches, AndMesh]]:
>> plot3d(Scaling = Constrained, sphere);
```

Exercise 11.3: The function `plotlib::implicitplot` covers the set of roots by small squares. In contrast, `plotlib::contourplot` tries to compute the solution points with with higher precision and to join them by a curve[3]. The following calls yield the desired graphics:

```
>> f := (x, y) -> ((x^2 + y^2)*(x^2 + y^2 -1)):
>> plotlib::implicitplot(f, -2..2, -2..2);
>> plotlib::contourplot([[x, y, f(x,y)], x = [-2 ,2],
                         y = [-2,2], Contours = [0],
                         Grid = [30, 30]]);
```

Exercise 14.1: With the following definition of the `postOutput` method of `Pref` the system prints an additional status line:

```
>> Pref::postOutput(
     proc()
     begin
       "bytes: " .
       expr2text(op(bytes(), 1)) . " (logical) / " .
       expr2text(op(bytes(), 2)) . " (physical)"
     end_proc):
>> DIGITS := 10: float(sum(1/i!, i = 0..100));

   2.718281828
   bytes: 836918 (logical) / 1005938 (physical)
```

---

[3] MuPAD versions beyond 1.4 implement an improved algorithm for `plotlib::implicitplot`, and `plotlib::contourplot` is obsolete. Moreover, the first argument of `plotlib::implicitplot` must be a MuPAD expression instead of a function. Consult the help page `?plotlib::implicitplot` for further information.

Exercise 15.1: We first generate the set S:

```
>> f := i -> ( (i^(5/2)+i^2-i^(1/2)-1) /
               (i^(5/2)+i^2+2*i^(3/2)+2*i+i^(1/2)+1)
             ):
>> S := {f(i) $ i=-1000..-2} union {f(i) $ i=0..1000}:
```

Then we apply **domtype** to all elements of the set to determine their domain types:

```
>> map(S, domtype);

  {DOM_INT, DOM_RAT, DOM_EXPR}
```

You see the explanation for this result by looking at some elements:

```
>> f(-2), f(0), f(1), f(2), f(3), f(4);
```

$$\frac{3\,I\,2^{1/2} + 3}{I\,2^{1/2} + 1}, -1, 0, \frac{3\,2^{1/2} + 3}{9\,2^{1/2} + 9}, \frac{8\,3^{1/2} + 8}{16\,3^{1/2} + 16}, 3/5$$

The function **normal** simplifies the expressions containing square roots:

```
>> map(%, normal);

  3, -1, 0, 1/3, 1/2, 3/5
```

Now we apply **normal** to all elements of the set before querying their data type:

```
>> map(S, domtype@normal);

  {DOM_INT, DOM_RAT}
```

Thus all numbers in S are indeed rational (in particular there are two integral values $f(0)=-1$ and $f(1)=0$). The reason is that $f(i)$ can be simplified to $(i-1)/(i+1)$:

```
>> normal(f(i) - (i - 1)/(i + 1));
```

    0

Exercise 15.2: We apply testtype(.,"sin") to each element of the list:

```
>> list := [sin(i*PI/200) $ i = 0..100]:
```

to find out, whether it is returned in the form sin(.). The following split command (Section 4.7) decomposes the list accordingly:

```
>> decomposition := split(list, testtype, "sin"):
```

In MuPAD 1.4, the system has simplified only 6 of the 101 sin calls:

```
>> map(decomposition, nops); decomposition[2];
```

    [95, 6, 0]

```
 --          1/2 1/2   1/2         1/2 1/2   1/2
 |       (2 - 2   )     2    (5 - 5   )       2
 | 0,  -------------, -------------------, ----,
 --          2                 4             2

     1/2     1/2    --
    (2   + 2)       |
   -------------, 1 |
         2          --
```

Exercise 15.3: You can use select (Section 4.7) to extract those elements that testtype identifies as positive integers. For example:

```
>> set := {-5, 2.3, 2, x, 1/3, 4}:
>> select(set, testtype, Type::PosInt);
```

    {2, 4}

Note that this selects only those objects that *are* positive integers, but not those that might *represent* positive integers, such as the identifier x in the above example. This is not possible with testtype. Instead you can use assume to set this property and query it via is:

```
>> assume(x, Type::PosInt):
>> select(set, is, Type::PosInt);

   {x, 2, 4}
```

Exercise 15.4: We construct the desired type specifier and employ it as follows:

```
>> T := Type::ListOf(Type::ListOf(
        Type::AnyType, 3, 3), 2, 2);

   ListOf(ListOf(AnyType, 3, 3), 2, 2)

>> testtype([[a, b, c], [1, 2, 3]], T),
   testtype([[a, b, c], [1, 2]], T);

   TRUE, FALSE
```

Exercise 17.1: Consider the conditions x<>1 and A and x=1 or A, respectively. After entering:

```
>> x := 1:
```

it is not possible to evaluate them due to the singularity in $x/(x-1)$:

```
>> x <> 1 and A;

   Error: Division by zero [_power]

>> x = 1 or A;

   Error: Division by zero [_power]
```

However, this is not a problem within an if statement since the Boolean evaluation of x<>1 and x=1 already tells us that x<>1 and A evaluates to FALSE and x=1 or A to TRUE, respectively:

```
>> (if x <> 1 and A then right else wrong end_if),
   (if x = 1 or A then right else wrong end_if);


   wrong, right
```

On the other hand, evaluation of the following if statement still produces an error, since it is necessary to evaluate A in order to determine the truth value of x=1 and A:

```
>> if x = 1 and A then right else wrong end_if;


   Error: Division by zero [_power]
```

Exercise 18.1: In analogy to the Airy example, we perform the following steps:

```
>> Ai := proc(x, a, b) begin
           if x = 0 then a else procname(args()) end_if
         end_proc:
>> Ai := func_env(Ai, NIL, NIL):
>> proc(f, x) local y, a, b; begin
     y := op(f, 1); a := op(f, 2); b := op(f, 3);
     Ai1(y, a, b)*diff(y, x)
   end_proc:
>> Ai := funcattr(Ai, "diff", %):
>> Ai1 := proc(x, a, b) begin
           if x = 0 then b else procname(args()) end_if
         end_proc:
>> proc(f) begin "Ai'(".expr2text(op(f)).")" end_proc:
>> Ai1 := func_env(Ai1, %, NIL):
>> proc(f, x) local y, a, b; begin
     y := op(f, 1); a := op(f, 2); b := op(f, 3);
     y*Ai(y, a, b)*diff(y, x)
   end_proc:
>> Ai1 := funcattr(Ai1, "diff", %):
```

Now we have:

```
>> diff(Ai(x, a, b), x, x);

   x Ai(x, a, b)

>> diff(Ai1(2*x + 3, a, b), x, x, x);
```

$$16 \text{ Ai'}(x*2 + 3, a, b) + 8 (2 x + 3)^2 \text{ Ai}(2 x + 3, a, b)$$

We compute the requested 10th derivative:

```
>> diff(Ai(x, 1, 0), x $ 10);
```

$$80 \text{ Ai'}(x, 1, 0) + 100 x^2 \text{ Ai}(x, 1, 0) + x^5 \text{ Ai}(x, 1, 0)$$

$$+ 20 x^3 \text{ Ai'}(x, 1, 0)$$

The first terms of the Taylor expansion around $x = 0$ are:

```
>> taylor(Ai(x, a, b), x = 0, 10);
```

$$\text{Ai}(0, a, b) + x \text{ Ai'}(0, a, b) + \frac{x^3 \text{ Ai}(0, a, b)}{6} +$$

$$\frac{x^4 \text{ Ai'}(0, a, b)}{12} + \frac{x^6 \text{ Ai}(0, a, b)}{180} + \frac{x^7 \text{ Ai'}(0, a, b)}{504}$$

$$+ \frac{x^9 \text{ Ai}(0, a, b)}{12960} + O(x^{10})$$

In MuPAD version 1.4, we have to enforce the evaluation of `Ai(0,a,b)` and `Ai'(0,a,b)`:

```
>> t := expr(%): t;
```

$$a + b x + \frac{a x^3}{6} + \frac{b x^4}{12} + \frac{a x^6}{180} + \frac{b x^7}{504} + \frac{a x^9}{12960}$$

Exercise 18.2: The following procedure evaluates the `Abs` function:

```
>> Abs := proc(x)
   begin
     if domtype(x) = DOM_INT or domtype(x) = DOM_RAT
       or domtype(x) = DOM_FLOAT
       then if x >= 0 then x else -x end_if;
       else procname(x);
     end_if
   end_proc:
```

We convert `Abs` to a function environment and supply a function producing the desired the screen output:

```
>> Abs := func_env(Abs,
                proc(f) begin
                   "|".expr2text(op(f))."|"
                end_proc,
                NIL):
```

Then we set the function attribute for differentiation:

```
>> Abs := funcattr(Abs, "diff",
                proc(f,x) begin
                   f/op(f)*diff(op(f),x)
                end_proc):
```

Now we have the following behavior:

```
>> Abs(-23.4), Abs(x), Abs(x^2 + y - z);

   23.4, |x|, |y + z*(-1) + x^2|
```

The **diff** attribute of the system function **abs** is implemented slightly differently:

```
>> diff(Abs(x^3), x), diff(abs(x^3), x);

   3 |x^3|                2
   -------, 3 abs(x)  sign(x)
      x
```

Exercise 18.3: We use **expr2text** (Section 4.11) to convert the integers passed as arguments to strings. Then we combine them, together with some slashes, via the concatenation operator ".":

```
>> date := proc(month, day, year) begin
              print(Unquoted, expr2text(month) . "/" .
                              expr2text(day) . "/" .
                              expr2text(year))
           end_proc:
```

Exercise 18.4: We present a solution using a **while** loop. The condition x mod 2 = 0 checks whether x is even:

```
>> f := proc(x) local i;
   begin
     i := 0;
     userinfo(2, "term " . expr2text(i) . ": " .
                 expr2text(x));
     while x <> 1 do
       if x mod 2 = 0 then x := x/2
       else x := 3*x+1 end_if;
       i := i + 1;
       userinfo(2, "term " . expr2text(i) . ": " .
                   expr2text(x))
     end_while;
     i
   end_proc:
```

```
>> f(4), f(1234), f(56789), f(123456789);
```

```
2, 132, 60, 177
```

If we set `setuserinfo(f,2)` (Section 14.2), then the `userinfo` command outputs all terms of the sequence until the procedure terminates:

```
>> setuserinfo(f, 2): f(4);
```

```
term 0: 4
term 1: 2
term 2: 1
```

```
2
```

If you do not believe in the $3x + 1$ conjecture, then you should insert a stopping condition for the index i to ensure termination.

Exercise 18.5: A recursive implementation based on the relation $\gcd(a, b) = \gcd(a \bmod b, b)$ leads to an infinite recursion: we have $a \bmod b \in \{0, 1, \ldots, b-1\}$, and hence

$$(a \bmod b) \bmod b = a \bmod b$$

in the next step. Thus the function gcd would always call itself recursively with the same arguments. However, a recursive call of the form $\gcd(a, b) = \gcd(b, a \bmod b)$ make sense. Since $a \bmod b < b$, the function calls itself recursively for decreasing values of the second argument, which finally becomes zero:

```
>> Gcd := proc(a, b) begin      /* recursive variant */
              if b = 0
                  then a
                  else Gcd(b, a mod b)
              end_if
          end_proc:
```

For large values of a and b, you may need to increase the value of the environment variable MAXDEPTH, if Gcd exhausts the valid recursion depth. The following iterative variant avoids this problem:

```
>> GCD := proc(a, b) local c;  /* iterative variant */
       begin
           while b <> 0 do
               c := a; a := b; b := c mod b
           end_while;
           a
       end_proc:
```

This yields:

```
>> a := 123456: b := 102880:
>> Gcd(a, b), GCD(a, b), igcd(a, b), gcd(a, b);

   20576, 20576, 20576, 20576
```

Exercise 18.6: In the following implementation we generate a shortened copy $Y = [x_1, \ldots, x_n]$ of $X = [x_0, \ldots, x_n]$ and compute the list of differences $[x_1 - x_0, \ldots, x_n - x_{n-1}]$ via zip and _subtract (note that _subtract(y,x)=y-x). Then we multiply each element of this list with the corresponding numerical value in the list $[f(x_0), f(x_1), \ldots]$. Finally the function _plus adds all elements of the resulting list:

$$[(x_1 - x_0) f(x_0), \ldots, (x_n - x_{n-1}) f(x_{n-1})]:$$

```
>> Quadrature := proc(f, X)
   local Y, distances, numericalValues, products;
   begin
       Y := X; unassign(Y[1]);
       distances := zip(Y, X, _subtract);
       numericalValues := map(X, float@f);
       products := zip(distances, numericalValues, _mult);
       _plus(op(products))
   end_proc:
```

In the following example, we use $n = 1000$ equidistant sample points in the interval $[0, 1]$:

```
>> f := x -> (x*exp(x)): n := 1000:
>> Quadrature(f, [i/n $ i = 0..n]);

   0.9986412288
```

This is a (crude) numerical approximation of $\int_0^1 x\, e^x\, dx\ (= 1)$.

Exercise 18.7: The specification of Newton requires that the first argument f be an *expression* and not a MuPAD function. Thus to compute the derivative, we first use indets to determine the unknown in f. We substitute a numerical value for the unknown to evaluate the iteration function $F(x) = x - f(x)/f'(x)$ at a point:

```
>> Newton := proc(f, x0, n)
     local vars, x, F, sequence, i;
     begin
       vars := indets(float(f)):
       if nops(vars) <> 1
         then error(
         "the function must contain exactly one unknown"
                   )
         else x := op(vars)
       end_if;
       F := x - f/diff(f,x); sequence := x0;
       for i from 1 to n do
           x0 := float(subs(F, x = x0));
           sequence := sequence, x0
       end_for;
       return(sequence)
     end_proc:
```

In the following example, Newton computes the first terms of a sequence rapidly converging to the solution $\sqrt{2}$:

```
>> Newton(x^2 - 2, 1, 6);
```

```
  1, 1.5, 1.416666666, 1.414215686, 1.414213562,

  1.414213562, 1.414213562
```

Exercise 18.8: The call numlib::g_adic(..,2) yields the binary expansion of an integer as a list of bits:

```
>> numlib::g_adic(7, 2), numlib::g_adic(16, 2);
```

$$[1, 1, 1], [0, 0, 0, 0, 1]$$

Instead of calling numlib::g_adic directly, our solution uses a subprocedure binary furnished with the option remember. This accelerates the computation notably since numlib::g_adic is called frequently with the same arguments. The call isSPoint([x,y]) returns TRUE, when the point specified by the list [x,y] is a Sierpinski point. To check this, the function multiplies the lists with the bits of the two coordinates. At those positions where both x and y have a 1 bit, multiplication yields a 1. In all other cases $0 \cdot 0$, $1 \cdot 0$, $0 \cdot 1$ the result is 0. If the list of products contains at least one 1, then the point is a Sierpinski point. We use select (Section 4.6) to extract the Sierpinski points from all points considered. Then we convert each entry [x,y] of the resulting list to the form point(x,y) via map. In this format we can directly pass the resulting list of graphics points to plot2d:

```
>> Sierpinski := proc(xmax, ymax)
   local binary, isSPoint, allPoints, i, j, SPoints;
   begin
     binary := proc(x) option remember; begin
                  numlib::g_adic(x, 2)
               end_proc;
     isSPoint := proc(Point) local x, y; begin
                   x := binary(Point[1]);
                   y := binary(Point[2]);
                   has(zip(x, y, _mult), 1)
                 end_proc;
     allPoints := [([i, j] $ i = 1..xmax) $ j = 1..ymax];
     SPoints := select(allPoints, isSPoint);
     SPoints := map(SPoints, point@op);
     plot2d(ForeGround = RGB::Black,
            BackGround = RGB::White,
            [Mode = List, SPoints])
   end_proc:
```

For xmax=ymax=100 you already obtain a quite appealing picture:

```
>> Sierpinski(100, 100);
```

Exercise 18.9: We present a recursive solution. For an expression formula(x1,x2,..) with the identifiers x1,x2,.. we collect the identifiers in the set x={x1,x2,..} by means of indets. Then we substitute TRUE and FALSE, respectively, for x1 (= op(x,1)), and call the procedure recursively with the arguments formula(TRUE,x2,x3,..) and formula(FALSE,x2,x3,..), respectively. In this way, we test all possible combinations of TRUE and FALSE for the identifiers until the expression can finally be simplified to TRUE or FALSE at the bottom of the recursion. This value is returned to the calling procedure. If at least one of the TRUE/FALSE combinations yields TRUE, then the procedure returns TRUE, indicating that the formula is satisfiable, and otherwise it returns FALSE.

```
>> satisfiable := proc(formula) local x;
   begin
     x := indets(formula);
     if x = {} then return(formula) end_if;
     return(satisfiable(subs(formula, op(x, 1) = TRUE))
         or satisfiable(subs(formula, op(x, 1) = FALSE)))
   end_proc:
```

If the number of identifiers in the input formula is n, then the recursion depth is at most n and the total number of recursive calls of the procedure is at most 2^n. We apply this procedure in two examples:

```
>> F1 := ((x and y) or (y or z)) and (not x) and y and z:
>> F2 := ((x and y) or (y or z)) and (not y) and (not z):
>> satisfiable(F1), satisfiable(not F1),
   satisfiable(F2), satisfiable(not F2);

   TRUE, TRUE, FALSE, TRUE
```

The call simplify(.,logic) (Section 9.2) simplifies logical formulae. Formula F2 can be simplified to false, no matter what the values of x, y, and z are:

```
>> simplify(F1, logic), simplify(F2, logic);

   not x and y and z, FALSE
```

Documentation and References

You find a survey of the currently available MuPAD documentation at the web site:

http://www.mupad.de

In MuPAD versions up to 1.4, the MuPAD manual:

[MuP 96] THE MuPAD GROUP. *MuPAD User's Manual.* Wiley–
Teubner, 1996.

is available in MuPAD's online help system. On a Windows system, you open it by choosing "Open Manual" in the "Help" Menu. Use the item "To Page" in the "Go" menu (or the "Page" button on a UNIX system) of the help window to navigate to an arbitrary page of the book. The table of contents on page 3 lists all available chapters. On UNIX platforms, the command ? opens the table of contents of the manual. On a Macintosh, you open the manual via ?manual.

By choosing "Helpindex" (which is an item in the "Targets" menu on Windows platforms) in the help window, you can navigate to "Additional Documentation", such as, for example,

[Oev 98] W. OEVEL. *MuPAD 1.4: an Overview.* 1998.
[Dre 97] K. DRESCHER. *Axioms, Categories and Domains.* Automath
Technical Report No. 1, 1997.
[DPS 97] K. DRESCHER, F. POSTEL AND T. SCHULZE. *Advanced
Demonstrations with MuPAD 1.4.* Automath Technical Report, 1997.
[Pos 97] F. POSTEL. *A Demonstration Tour through MuPAD 1.4.* Automath Technical Report, 1997.

The MuPAD Quick Reference [Oev 98] lists all data types, functions, and libraries of MuPAD version 1.4, and provides a survey of its functionality. The documents [Pos 97] and [DPS 97] give demonstrations of MuPAD 1.4. You can access these interactive texts directly from within a MuPAD session via ?demo and ?advdemo, respectively.

You find descriptions of the MuPAD libraries by choosing "Helpindex" (from the "Targets" menu on Windows platforms) in the help window, and then using the link "Library Packages". There you find links to the individual libraries such as, for example, Dom (the library for preinstalled data types). The corresponding documentation [Dre 95] contains a concise description of all domains provided by Dom. In a MuPAD session the command ?Dom directly opens this document. Moreover, you can access the description of individual data structures from this document, such as Dom::Matrix, directly through the call ?Dom::Matrix. Another example is the documentation [Pos 98] for the linalg package (linear algebra), which can be requested directly via ?linalg:

[Dre 95]   K. DRESCHER. *Domain–Constructors*. Automath Technical Report No. 2, 1995.

[Pos 98]   F. POSTEL. *The Linear Algebra Package "linalg"*. Automath Technical Report No. 9, 1998.

It is not possible to print these documents and help pages from within the help system. You can download the compressed postscript documents from the above MuPAD web site.

In addition to the MuPAD documentation, we recommend the following books about computer algebra in general and the underlying algorithms:

[AHU 74]   A.V. AHO, J.E. HOPCROFT AND J.D. ULLMAN. *The Design and Analysis of Computer Algorithms*. Addison–Wesley, 1974.

[DTS 93]   J.H. DAVENPORT, E. TOURNIER AND Y. SIRET. *Computer Algebra: Systems and Algorithms for Algebraic Computation*. Academic Press, 1993.

[GCL 92]   K.O. GEDDES, S.R. CZAPOR AND G. LABAHN. *Algorithms for Computer Algebra*. Kluwer, 1992.

[GG 99]   J. VON ZUR GATHEN AND J. GERHARD. *Modern Computer Algebra*. Cambridge University Press, 1999.

[Hec 93]   A. HECK. *Introduction to Maple*. Springer, 1993.

Index

Printed in the United States
By Bookmasters